Structural Systems
for Tall Buildings

Library of Congress Cataloging-in-Publication Data

Structural systems for tall buildings / Council on Tall Buildings
   and Urban Habitat, Committee 3 ; Contributors, I.D. Bennetts...[et
   al.] ; editorial group, Ryszard M. Kowalczyk, Robert Sinn, Max B.
   Kilmister.
         p.   cm.
   Includes bibliographical references and index.
   ISBN 0-07-012541-4
      1. Tall buildings—Design and construction.   2. Structural
   engineering.   I. Bennetts, I. D.   II. Kowaczyk, Ryszard M.
   III. Sinn, Robert.   IV. Kilmister, Max B.   V. Council on Tall
   Buildings and Urban Habitat.   Committee 3.
   TH1611.S78   1995
   690—dc20   94-38031
                                                           CIP

1 2 3 4 5 6 7 8 9 0   DOC/DOC   9 0 9 8 7 6 5 4

ISBN 0-07-012541-4

*For the Council on Tall Buildings, Lynn S. Beedle is the Editor-in-Chief and
Dolores B. Rice is the Managing Editor.*

*For McGraw-Hill, the sponsoring editor was Joel Stein, the editing supervisor
was David E. Fogarty, and the production supervisor was Donald Schmidt.
This book was set in Times Roman by the Universities Press (Belfast) Ltd.*

This book is printed on acid-free paper.

Council on Tall Buildings and Urban Habitat

Contributors

Office of Irwin G. Cantor, P.C., New York
H.K. Cheng & Partners Ltd, Hong Kong
Douglas Specialist Contractors Ltd., Aldridge
Hart Consultant Group, Santa Monica
The George Hyman Construction Co.,
 Bethesda
Ingenieurburo Muller Marl GmbH, Marl
Institute Sultan Iskandar, Johor
INTEMAC, Madrid
JHS Construcao e Planejamento Ltd., Sao
 Paulo
Johnson Fain and Pereira Assoc., Los Angeles
The Kling-Lindquist Partnership, Inc.
 Philadelphia
LeMessurier Consultants Inc., Cambridge

Lim Consultants, Inc., Cambridge
Meinhardt Australia Pty. Ltd., Melbourne
Meinhardt (HK) Ltd., Hong Kong
Mueser Rutledge Consulting Engineers,
 New York
Obayashi Corporation, Tokyo
OTEP International, SA, Madrid
Charles Pankow Builders, Inc., Altadena
Projest SA Empreendimentos e Servicos
 Tecnicos, Rio de Janeiro
PSM International, Chicago
Skilling Ward Magnusson Barkshire Inc.,
 Seattle
Tooley & Company, Los Angeles
Nabih Youssef and Associates, Los Angeles

Contributing Participants

Advanced Structural Concepts, Denver
Adviesbureau Voor Bouwtechniek BV, Arnhem
American Institute of Steel Construction, Chicago
Anglo American Property Services (Pty) Ltd., Johannesburg
Architectural Services Dept., Hong Kong
Atelier D'Architecture, de Genval, Genval
Australian Institute of Steel Construction, Milsons Point
B.C.V. Progetti S.r.l., Milano
W.S. Bellows Construction Corp., Houston
Alfred Benesch & Co., Chicago
Bolsa de Imoveis Est. Sao Paulo, S.A., Sao Paulo
Bornhorst & Ward Pty. Ltd., Spring Hill
Boundary Layer Wind Tunnel Laboratory (U. Western Ontario), London
Bovis Limited, London
Brandow & Johnston Associates, Los Angeles
Brooke Hillier Parker, Hong Kong
Buildings & Data, S.A., Brussels
CBM Engineers Inc., Houston
Cermak Peterka Petersen, Inc., Fort Collins
CMA Architects & Engineers, San Juan
Construction Consulting Laboratory, Dallas
Crane Fulview Door Co., Lake Bluff
Crone & Associates Pty. Ltd., Sydney
Davis Langdon & Everett, London
DeSimone, Chaplin & Dobryn Inc., New York
Dodd Pacific Engineering, Inc., Seattle
Fujikawa Johnson and Associates, Chicago
Gutteridge Haskins & Davey Pty Ltd, Sydney
Haines Lundberg Waehler International, New York
Hayakawa Associates, Los Angeles
Healthy Buildings International Inc., Fairfax
Hellmuth, Obata & Kassabaum, Inc., San Francisco
International Iron & Steel Institute, Brussels
Irwin Johnston and Partners, Sydney
Jatocret, S.A., Rio de Janeiro
J.A. Jones Construction Co., Charlotte
Keating Mann Jernigan Rottet, Los Angeles
KPFF Consulting Engineers, Seattle
Lend Lease Design Group Ltd., Sydney

Lerch Bates & Associates Ltd., Surrey
Stanley D. Lindsey & Assoc., Nashville
Martin & Bravo, Inc., Honolulu
Martin, Middlebrook & Louie, San Francisco
Enrique Martinez-Romero, S.A., Mexico
Mitchell McFarlane Brentnall & Partners Intl. Ltd.,
 Hong Kong
Mitsubishi Estate Co.,Ltd., Tokyo
Moh and Associates, Inc., Taipei
Morse Diesel International, New York
Multiplex Constructions (NSW) Pty. Ltd., Sydney
Nihon Sekkei, U.S.A., Ltd., Los Angeles
Nikken Sekkei, Ltd., Tokyo
Norman Disney & Young, Brisbane
Pacific Atlas Development Corp., Los Angeles
Peddle Thorp Australia Pty. Ltd., Brisbane
Park Tower Group, New York
Cesar Pelli & Associates, New York
Perkins & Will, Chicago
Rahulan Zain Associates, Kuala Lumpur
RFB Consulting Architects, Johannesburg
Rosenwasser Grossman Cons. Engrs., PC, New York
Emery Roth & Sons Intl. Inc., New York
Rowan Williams Davies & Irwin Inc., Guelph
Sepakat Setia Perunding (Sdn.) Bhd., Kuala Lumpur
Sergen Servicos Gerais de Engenharia S.A., Rio de
 Janeiro
Severud Associates Cons. Engrs., New York
SOBRENCO, S.A., Rio de Janeiro
South African Institute of Steel Construction, Johannesburg
Steel Reinforcement Institute of Australia, Sydney
STS Consultants Ltd., Northbrook
Studio Finzi, Nova E Castellani, Milano
Taylor Thomson Whitting Pty Ltd, St. Leonards
B.A. Vavaroutas & Associates, Athens
VIPAC Engineers & Scientists Ltd, Melbourne
Wargon Chapman Partners, Sydney
Weidlinger Associates, New York
Woodward-Clyde Consultants, New York
Yapi Merkezi Inc., Istanbul

Other Books in the Tall Buildings and Urban Environment Series

Structural Systems for Tall Buildings

Council on Tall Buildings and Urban Habitat

Committee 3

CONTRIBUTORS
I.D. Bennetts
Joseph Burns
Brian Cavill
P.H. Dayawansa
Eiji Fukuzawa
Max B. Kilmister
Ryszard M. Kowalczyk
Owen Martin
William Melbourne
Seiichi Muramatsu
T. Okoshi
Ahmad Rahimian
Thomas Scarangello
Robert Sinn
Richard Tomasetti
A. Yamaki

Editorial Group

Ryszard M. Kowalczyk, Chairman
Robert Sinn, Vice-Chairman
Max B. Kilmister, Editor

McGraw-Hill, Inc.

New York San Francisco Washington, D.C. Auckland Bogotá
Caracas Lisbon London Madrid Mexico City Milan
Montreal New Delhi San Juan Singapore
Sydney Tokyo Toronto

ACKNOWLEDGMENT OF CONTRIBUTIONS

This Monograph was prepared by Committee 3 (Structural Systems) of the Council on Tall Buildings and Urban Habitat as part of the *Tall Buildings and Urban Environment* Series. The editorial group was Ryszard M. Kowalczyk, chairman; Robert Sinn, vice-chairman; and Max B. Kilmister, editor.

Special acknowledgment is due those individuals whose manuscripts formed the major contribution to the chapters in this volume. These individuals and the chapters or sections to which they contributed are:

Chapter 1: Editorial Group
Chapter 2: Editorial Group
Section 3.1: Editorial Group
Section 3.2: Brian Cavill
Section 4.1: Eiji Fukuzawa
Section 4.1: Seiichi Muramatsu
Section 4.1: Ahmad Rahimian
Section 4.2: Owen Martin
Section 4.3: T. Okoshi

Section 4.3: Thomas Scarangello
Section 4.3: Richard Tomasetti
Section 4.3: A. Yamaki
Section 4.4: Editorial Group
Section 4.5: Editorial Group
Section 5.1: William Melbourne
Section 5.2: I. D. Bennetts
Section 5.2: P. H. Dayawansa
Chapter 6: Joseph Burns

Project Descriptions were contributed by:

The Office of Irwin Cantor
CBM Engineers, Inc.
Ellisor and Tanner, Inc.
Kajima Design, Inc.
King/Guinn Associates
LeMessurier Consultants, Inc.
Leslie E. Robertson Associates
Nihon Sekkei, Inc.
Ove Arup & Parners

Paulus, Sokolowski, and Sartor, Inc.
Perkins and Will
Robert Rosenwasser Associates
Severud Associates
Shimizu Corporation
Skidmore, Owings and Merrill
Skilling Ward Magnusson Barkshire, Inc.
Thornton-Tomasetti Engineers
Walter P. Moore and Associates

COMMITTEE MEMBERS

Herbert F. Adigun, Mir M. Ali, Luis Guillermo Aycardi, Prabodh V. Banavalkar, Bob A. Beckner, Charles L. Beckner, George E. Brandow, John F. Brotchie, Robert J. Brungraber, Yu D. Bychenkov, Peter W. Chen, Ching-Churn Chern, Pavel Cizek, Andrew Davids, John DeBremaeker, Dirk Dicke, Robert O. Disque, Richard Dziewolski, Ehua Fang, Alexander W. Fattaleh, James G. Forbes, Robert J. Hansen, Robert D. Hansen, Toshiharu Hisatoku, Arne Johnson, Michael Kavyrchine, Max B. Kilmister (editor), Gert F. Konig, Ryszard M. Kowalczyk (chairman), Juraj Kozak, Monsieur G. Lacombe, Siegfried Liphardt, Miguel A. Macias-Rendon, Owen Martin, Jaime Mason, N. G. Matkov, Gerardo G. Mayor, Leonard R. Middleton, Jaime Munoz-Duque, Jacques Nasser, Anthony F. Nassetta, Fujio Nishikawa, Alexis Ostapenko, Z. Pawlowski, M. V. Posokhin, Peter Y. S. Pun, Werner Quasebarth, Govidan Rahulan, Anthony Fracis Raper, Satwant S. Rihal, Leslie E. Robertson, Wolfgang Schueller, Duiliu Sfintesco, Robert Sinn (vice-chairman), Ramiro A. Sofronie, A. G. Sokolov, Etsuro Suzuki, Bungale S. Taranath, A. R. Toakley, Kenneth W. Wan, Morden S. Yolles, Nabih F. G. Youssef, Stefan Zaczek.

GROUP LEADERS

The committee on Structural Systems is part of Group SC of the Council, "Systems and Concepts." The leaders are:

James G. Forbes, Chairman
Joseph P. Colaco, Vice-Chairman
Henry J. Cowan, Editor

Foreword

This volume is one of a series of Monographs prepared under the aegis of the Council on Tall Buildings and Urban Habitat, a series that is aimed at documenting the state of the art of the planning, design, construction, and operation of tall buildings as well as their interaction with the urban environment.

The present series is built upon an original set of five Monographs published by the American Society of Civil Engineers, as follows:

Volume PC: Planning and Environmental Criteria for Tall Buildings

Volume SC: Tall Building Systems and Concepts

Volume CL: Tall Building Criteria and Loading

Volume SB: Structural Design of Tall Steel Buildings

Volume CB: Structural Design of Tall Concrete and Masonry Buildings

Following the publication of a number of updates to these volumes, it was decided by the Steering Group of the Council to develop a new series. It would be based on the original effort but would focus more strongly on the individual topical committees rather than the groups. This would do two things. It would free the Council committees from restraints as to length. Also it would permit material on a given topic to reach the public more quickly.

The result was the *Tall Buildings and Urban Environment* series, being published by McGraw-Hill, Inc., New York. The present Monograph joins six others, the first of which was released in 1992:

Cast-in-Place Concrete in Tall Building Design and Construction

Cladding

Building Design for Handicapped and Aged Persons

Fire Safety in Tall Buildings

Semi-Rigid Connections in Steel Frames

Cold-Formed Steel in Tall Buildings

This particular Monograph was prepared by the Council's Committee 3, Structural Systems. Its earlier treatment was a part of Volume SC. It dealt with the many issues relating to tall building structural systems when it was published in 1980. The committee decided that a volume featuring case studies of many of the most important buildings of the last two decades would provide professionals with some interesting comparisons of how and why structural systems were chosen. The result of the committee's efforts is this Monograph. It provides case studies of tall buildings from Japan, the United States, Malaysia, Australia, New Zealand, Hong Kong, Spain, and Singapore. This unique international survey examines the myriad of architectural, engineering, and construction issues that must be taken into account in designing tall building structural systems.

The Monograph Concept

The Monograph series is prepared for those who plan, design, construct, or operate tall buildings, and who need the latest information as a basis for judgment decisions. It includes a summary and condensation of recent developments for design use, it provides a major reference source to recent literature, and it identifies needed research.

The Council's Monograph series is not intended to serve as a primer. Its function is to communicate to all knowledgeable persons in the various fields of expertise the state of the art and the most advanced knowledge in those fields. Our message has more to do with reporting trends and general approaches rather than with detailed applications. It aims to provide adequate information for experienced professionals confronted with their first high rise, as well as opening new vistas to those who have been involved with them in the past. It aims at an international scope and interdisciplinary treatment.

The Monograph series was not designed to cover topics that apply to all buildings in general. However, if a subject has application to *all* buildings, but also is particularly important for a *tall* building, then the objective has been to treat that topic.

Direct contributions to this Monograph have come from many sources. Much of the material has been prepared by those in actual practice as well as by those in the academic sector. The Council has seen considerable benefit accrue from the mix of professions, and this is no less true in the Monograph series itself.

Tall Buildings

A tall building is not defined by its height or number of stories. The important criterion is whether or not the design is influenced by some aspect of "tallness." It is a building in which tallness strongly influences planning, design, construction, and use. It is a building whose height creates conditions different from those that exist in "common" buildings of a certain region and period.

The Council

The Council is an international group sponsored by engineering, architectural, construction, and planning professionals throughout the world, an organization that was established to study and report on all aspects of the planning, design, construction, and operation of tall buildings.

The sponsoring societies of the Council are the American Institute of Architects (AIA), American Society of Civil Engineers (ASCE), American Planning Association (APA), American Society of Interior Designers (ASID), American Society of Heating, Refrigerating, and Air-Conditioning Engineers (ASHRAE), International Association for Bridge and Structural Engineering (IABSE), International Union of Architects (UIA), Japan Structural Consultants Association (JSCA), Urban Land Institute (ULI), and International Federation of Interior Designers (IFI).

The Council is concerned not only with the building itself but also with the role of tall buildings in the urban environment and their impact thereon. Such a concern also involves a systematic study of the whole problem of providing adequate space for life and work, considering not only technological factors, but social and cultural aspects as well.

Nomenclature and Other Details

The general guideline for units is to use SI units first, followed by American units in parentheses, and also metric when necessary. A conversion table for units is supplied at the end of the volume. A glossary of terms and a list of symbols also appear at the end of the volume.

The spelling was agreed at the outset to be "American" English.

The relevant references and bibliography will be found at the end of each chapter. A composite list of all references appears at the end of the volume.

From the start, the Tall Building Monograph series has been the prime focus of the Council's activity, and it is intended that its periodic revision and the implementation of its ideas and recommendations should be a continuing activity on both the national and the international levels. Readers who find that a particular topic needs further treatment are invited to bring it to our attention.

Acknowledgments

This work would not have been possible but for the early financial support of the National Science Foundation, which supported the program out of which this Monograph developed. More recently the major financial support has been from the organizational members, identified in earlier pages of this Monograph, as well as from many individuals members. Their confidence is appreciated.

All those who had a role in the authorship of the volume are identified in the acknowledgment page that follows the title page. Especially important are the contributors whose papers formed the essential first drafts—the starting point.

The primary conceptual and editing work was in the hands of the leaders of the Council's Committee 3, Structural Systems. The Chairman is Ryszard M. Kowalczyk of Universidade da Beira Interior, Covilha, Portugal. The Vice-Chairman is Robert Sinn of Skidmore, Owings and Merrill, Chicago, Illinois, USA. Comprehensive editing was the responsibility of Max B. Kilmister of Connell Wagner Rankine and Hill, Brisbane, Australia.

Overall guidance was provided by the Group Leaders: James G. Forbes of Irwin Johnston and Partners, Milsons Point, Australia; Joseph P. Colaco of CBM Engineers, Inc., Houston, Texas, USA; and Henry J. Cowan of the University of Sydney, Sydney, Australia.

Lynn S. Beedle
Editor-in-Chief

Lehigh University   *Dolores B. Rice*
Bethlehem, Pennsylvania   *Editor*
1994

Preface

Although tall buildings are generally considered to be a product of the modern industrialized world, inherent human desire to build skyward is nearly as old as human civilization. The ancient pyramids of Giza in Egypt, the Mayan temples in Tikal, Guatamala, and the Kutab Minar in India are just a few examples eternally bearing witness to this instinct. Skyscrapers in the modern sense began to appear over a century ago; however, it was only after World War II that rapid urbanization and population growth created the need for the construction of tall buildings.

The dominant impact of tall buildings on urban landscapes has tended to invite controversy, particularly in cities with older historic structures. The skyscraper silhouette has transformed and shaped the skylines of many cities, thereby creating the most characteristic and symbolic testaments to the cities' wealth and their inhabitants' collective ambitions.

The ordinary observer recognizes the tall building primarily with respect to its exterior architectural enclosure. This is only natural, as when we consider the great pyramids of Egypt our overriding image is of their characteristic shape. It is only recently that we have begun to realize the creativity and colossal effort expended by these ancient people to erect these structures in the desert at that time. So it is with the modern skyscraper. The overall spatial form as well as the intricate detailing of the cladding systems are crucial in defining the architectural expression and in placing the tower within the overall urban environment. The aim of this Monograph, however, is to have a look under the outer covering of the building to reveal the structural skeleton as well as to provide historical knowledge documenting the design and construction techniques used to realize these monuments in today's world.

This Monograph is therefore dedicated to the structural systems for tall buildings: their evolution and historical development as well as the variety of solutions engendered to allow the tower to be realized safely and efficiently. As in the past, new achievements in material science, computer-aided design, and construction technology have opened paths toward more sophisticated and elegant structural systems for tall buildings. The structural system organization chosen for a particular project determines the fundamental properties of the overall building, the behavior under imposed loads, its safety, and often may have a dramatic impact on the architectural design. The intent of this volume is to demonstrate the characteristic features of many outstanding system forms while documenting the factors leading to their selection for projects actually realized.

The structural systems for high-rise buildings are constantly evolving and at no time can be described as a completed whole. Every month new buildings are being designed and created, new projects conceived, and new schemes applied. Nevertheless, we hope it is worthwhile to present the current state of the art while being aware that progress in systems development is ongoing.

The planning for this Monograph began soon after the decision was made by the Council to expand the chapters of the original Monograph into separate volumes. The concept of a volume based on a survey of some of the most innovative examples of tall building structural systems contributed by leading engineers and design firms of the profession was conceived during the committee workship in Hong Kong in 1990. It was only after establishing the editorial leadership for the work that the volume began to take form, with the scope and content of the book finalized. At this time a building data form was prepared for collecting the most essential information concerning the structural design of the buildings included herein. The surveys were initiated and the responses compiled by Max Kilmister. This material represents the core of the completed book and the vast majority of the work. Bob Sinn then assembled all of the "loose ends" of the compilation in the summer of 1993 in order to finish the completed volume in time for publication.

The Monograph as a whole is a product of extensive teamwork. Sincere thanks go to all of the contributors who offered their valuable time to share their experience with the readers. It is around this information that the entire work is constructed. We hope that the information included may be presented to a broad professional audience. This exchange of information is one of the tenets of the Council and is in fact a condition for progress in the design of tall buildings.

Supporting information for Chapter 5 from Drs. B. J. Vickery, J. D. Holmes, and J. C. K. Cheung is gratefully acknowledged, as is the Australian Research Grants Commission for its support of the fundamental research.

As mentioned, we are aware that everyday progress is made in the field of structural engineering for high-rise buildings. The committee is already thinking about expanding and updating this volume. We urge all readers to enrich and complement this work by writing the Council or joining the committee.

Finally, we would like to express our appreciation to Dr. Lynn Beedle, who encouraged us to prepare this work and who advised and supported the effort. We dedicate this book to him.

Ryszard M. Kowalczyk
Chairman

Robert Sinn
Vice-Chairman

Max B. Kilmister
Editor

Contents

Structural Systems
for Tall Buildings

1

Introduction

Structural systems for tall buildings have undergone a dramatic evolution throughout the previous decade and into the 1990s. Developments in structural system form and organization have historically been realized as a response to as well as an impetus toward emerging architectural trends in high-rise building design. At the time of publication of the initial Council Monograph *Tall Building Systems and Concepts* in 1980, international style and modernist high-rise designs, characterized by prismatic, repetitive vertical geometries and flat-topped roofs, were predominant (Council on Tall Buildings, Group SC 1980). The development of the prototype tubular systems for tall buildings was indeed predicated upon an overall building form of constant or smoothly varying profile. A representative office building project from the period is shown in Fig. 1.1. The rigid discipline of the exterior tower form has since been replaced in many cases by the highly articulated vertical modulations of the building envelope characteristic of eclectic postmodern, deconstructivist, and neohistorical high-rise expressions (Fig. 1.2). This general discontinuity and erosion of the exterior facade has led to a new generation of tall building structural systems that respond to the more flexible and idiosyncratic requirements of an increasingly varied architectural aesthetic. Innovative structural systems involving megaframes, interior super-diagonally braced frames, hybrid steel and high-strength concrete core and outrigger systems, artificially damped structures, and spine structures are among the compositions which represent a step in the development of structural systems for high-rise buildings. This Monograph seeks to further the placement of some of the most exciting and unique forms for today's tall building structures into the overall tall building system hierarchy.

One of the fundamental goals of the Council has been to continually develop a tall buildings database. The members of Committee SC-3, Structural Systems, decided that rather than being a collection of papers or a general survey of tall building structural systems, the Monograph would be organized with respect to such a database-type format of structural and project information on actual building projects. The committee therefore requested detailed information from engineers in the profession, regarding the structural design of some of the most innovative high-rise projects throughout the world. An enthusiastic response from the structural engineering community provided very specific engineering information such as wind and seismic loadings, dynamic properties, materials, and systems for a wide range of international high-rise projects, both completed and in proposal stage, which are compiled in this single work. These comprehensive data are the primary focus of this Monograph and should

1

be of interest and value to practicing engineers and architects as well as other tall building enthusiasts.

This Monograph is organized into six chapters. A general introduction to the classification of tall building structural systems is found in Chapter 2. The section begins to define the parameters and characteristics for which tall building systems are evaluated. Tall building floor systems are discussed in Chapter 3, which includes recent

Fig. 1.1   Quaker Oats Tower, Chicago, Illinois, Completed 1984. (*Courtesy: Skidmore Owings & Merrill.*)

Fig. 1.2   NBC Tower, Chicago, Illinois, Completed 1991. (*Courtesy: Skidmore Owings & Merrill.*)

developments in posttensioned concrete floor systems for high-rise construction in Australia. Structural systems for tall buildings have historically been grouped with respect to their ability to resist lateral loads effectively. Therefore Chapter 4, "Lateral Load Resisting Systems," forms the core of the work, with system descriptions for over 50 projects. The projects are arranged within five basic subclassifications for lateral load resistance with generally increasing efficiency and application for taller buildings: braced frame and moment resisting frame systems, shear wall systems, core and outrigger systems, tubular systems, and hybrid systems. Each subsection is preceded by a general introduction outlining the system forms, limitations, advantages, and applications. Chapter 5 discusses special topics in high-rise building structural systems. It presents information concerning the developing topics of wind-induced motions and fire protection of structural members in tall buildings. The concluding Chapter 6, in dealing with systems for the future, presents examples of projects on the drawing board and proposals which represent innovative state-of-the-art structural designs for tall buildings.

1.1 CONDENSED REFERENCES/BIBLIOGRAPHY

Council on Tall Buildings, Group SC 1980, *Tall Building Systems and Concepts*.

2

Classification of Tall Building Structural Systems

The Council definition of a tall building defines the unique nature of the high-rise project: "A building whose height creates different conditions in the design, construction, and use than those that exist in common buildings of a certain region and period." For the practicing structural engineer, the cataloging of structural systems for tall buildings has historically recognized the primary importance of the system to resist lateral loads. The progression of lateral load resisting schemes from elemental beam and column assemblages toward the notion of an equivalent vertical cantilever is fundamental to any structural systems methodology.

In 1965 Fazlur Khan (1966) recognized that this hierarchy of system forms could be roughly categorized with respect to relative effectiveness in resisting lateral loads (Fig. 2.1). At one end of the spectrum are the moment resisting frames, which are efficient for buildings in the range of 20 to 30 stories; at the other end is the generation of tubular systems with high cantilever efficiency. With the endpoints defined, other systems were placed with the idea that the application of any particular form is economical only over a limited range of building heights. The system charts were updated periodically as new systems were developed and improvements in materials and analysis techniques evolved.

Alternatively, the classification process could be based on certain engineering and systems criteria which define both the physical as well as the design aspects of the building:

- Material
  Steel
  Concrete
  Composite
- Gravity load resisting systems
  Floor framing (beams, slabs)
  Columns

5

Trusses

Foundations

- Lateral load resisting systems

Walls

Frames

Trusses

Diaphragms

- Type and magnitude of lateral loads

Wind

Seismic

- Strength and serviceability requirements

Drift

Acceleration

Ductility

In 1984 the Council attempted to develop a rigorous methodology for the cataloging of tall buildings with respect to their structural systems (Falconer and Beedle, 1984). The classification scheme involves four distinct levels of framing-oriented division: primary framing system, bracing subsystem, floor framing, and configuration

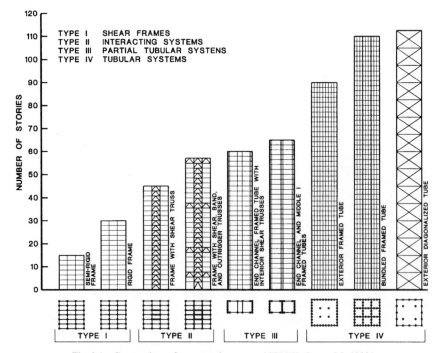

Fig. 2.1   Comparison of structural systems. (*CTBUH, Group SC, 1980.*)

and load transfer. These levels are further broken down into subgroups and discrete systems (Fig. 2.2). This format allows for the consistent and specific identification and documentation of tall buildings and their systems, the overriding goal being to achieve a comprehensive worldwide survey of the *performance* of buildings in the high-rise environment.

While any cataloging scheme must address the preeminent focus on lateral load resistance, the load-carrying function of the tall building subsystems is rarely independent. The most efficient high-rise systems fully engage vertical gravity load resisting elements in the lateral load subsystem in order to reduce the overall structural premium for resisting lateral loads. Some degree of independence is generally recognized between the *floor framing systems* and the *lateral load resisting systems,* although the integration of these subassemblies into the overall structural organization is crucial.

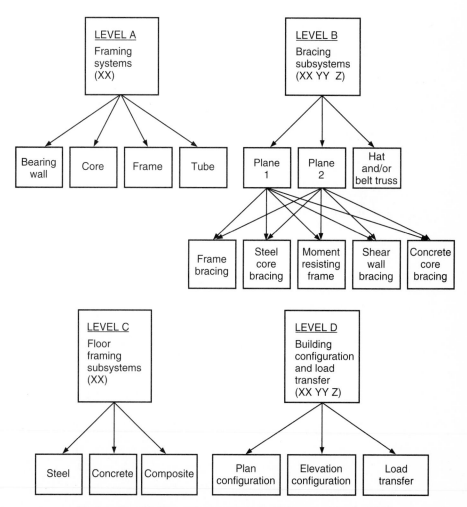

Fig. 2.2   Classification of structural systems. (*Falconer and Beedle, 1984.*)

This Monograph therefore divides the discussion of tall building structural systems into the subsystems mentioned.

2.1 CONDENSED REFERENCES/BIBLIOGRAPHY

Falconer and Beedle 1984, *Classification of Tall Building Systems.*
Khan 1966, *Optimization of Building Structures.*

3

Tall Building
Floor Systems

3.1 COMPOSITE STEEL FLOOR SYSTEMS

Composite floor systems typically involve simply supported structural steel beams, joists, girders, or trusses linked via shear connectors with a concrete floor slab to form an effective T-beam flexural member resisting primarily gravity loads. The versatility of the system results from the inherent strength of the concrete floor component in compression and the tensile strength and spannability of the steel member. Composite floor systems are advantageous because of reduced material costs, reduced labor due to prefabrication, faster construction times, simple and repetitive connection details, reduced structural depths and consequent efficient use of interstitial ceiling space, and reduced building mass in zones of heavy seismic activity. The composite floor system slab element can be formed by a flat-soffit reinforced concrete slab, precast concrete planks or floor panels with or without a cast-in-place topping slab, or a metal steel deck, either composite or noncomposite (Fig. 3.1). When a composite floor framing member is combined with a composite metal deck and a concrete floor slab, an extremely efficient system is formed. The composite action of the beam or truss element is due to shear studs welded directly through the metal deck, whereas the composite action of the metal deck results from side embossments incorporated into the steel sheet profile. The slab and beam arrangement typical in composite floor systems produces a rigid horizontal diaphragm, providing stability to the overall building system while distributing wind and seismic shears to the lateral load resisting system elements.

1 Composite Beams and Girders

Steel and concrete composite beams may be formed either by completely encasing a steel member in concrete, with the composite action depending on the natural bond caused by the chemical adhesion and mechanical friction between steel and concrete, or by connecting the concrete floor to the top flange of the steel framing member through shear connectors (Fig. 3.1). The concrete-encased composite steel beam was common prior to the development of sprayed-on cementitious and board or batt type fireproofing materials, which economically replaced the heavy formed concrete insulation on the steel beam. Today the most common arrangement found in composite

9

floor systems is a rolled or built-up steel beam connected to a formed steel deck and concrete slab. The metal deck typically spans unshored between steel members while also providing a working platform for steel erection. The metal deck slab may be oriented parallel or perpendicular to the composite beam span and may itself be either composite or noncomposite (form deck). Figure 3.2 shows a typical office building floor that is framed in composite steel beams.

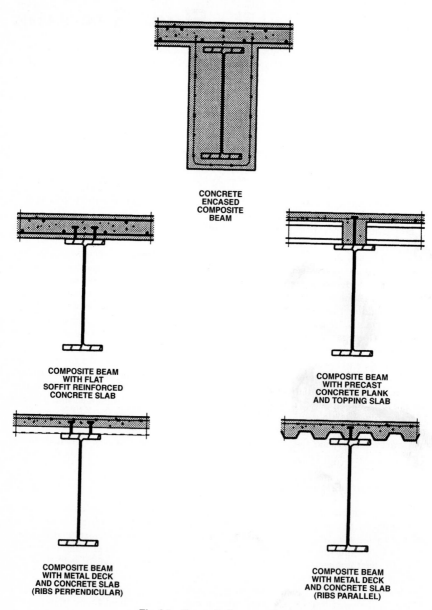

Fig. 3.1   Composite beam systems.

In composite beam design, the stress distribution at working loads across the composite section is shown schematically in Fig. 3.3. As the top flange of the steel section is normally quite near the neutral axis and consequently lightly stressed, a number of built-up or hybrid composite beam schemes have been formulated in an attempt to use the structural steel material more efficiently (Fig. 3.4). Hybrid beams fabricated from ASTM A36 grade top flange steel and 345-MPa (50-ksi)-yield bottom flange steel have been used. Also, built-up composite beam schemes or tapered flange beams are possible. In all of these cases, however, the increased fabrication costs must be evaluated, which tend to offset the relative material efficiency. In addition, a relatively wide and thick-gauge top flange must be provided for proper and effective shear stud installation.

A prismatic composite steel beam has two fundamental disadvantages over other types of composite floor framing types. (1) The member must be designed for the maximum bending moment near midspan and thus is often understressed near the sup-

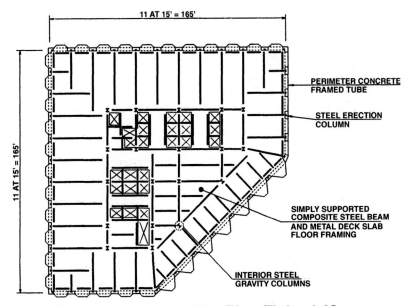

Fig. 3.2   Three First National Plaza, Chicago, Illinois, typical floor.

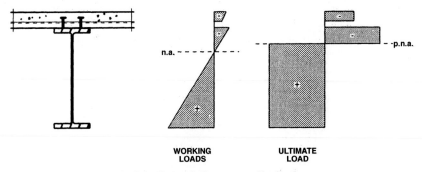

Fig. 3.3   Composite beam stress distribution.

ports, and (2) building-services ductwork and piping must pass beneath the beam, or the beam must be provided with web penetrations (normally reinforced with plates or angles leading to higher fabrication costs) to allow access for this equipment. For this reason, a number of composite girder forms allowing the free passage of mechanical ducts and related services through the depth of the girder have been developed. They include tapered and dapped girders, castellated beams, and stub girder systems (Fig. 3.5). As the tapered girders are completely fabricated from plate elements or cut from rolled shapes, these composite members are frequently hybrid, with the top flange designed in lower-strength steel. Applications of tapered composite girders to office building construction are limited since the main mechanical duct loop normally runs through the center of the lease span rather than at each end. The castellated composite beam is formed from a single rolled wide-flange steel beam cut and then reassembled by welding with the resulting increased depth and hexagonal openings. These members are available in standard shapes by serial size and are quite common in the United Kingdom and the rest of Europe. Use in the United States is limited due to the increased fabrication cost and the fact that the standard castellated openings are not large enough to accommodate the large mechanical ductwork common in modern high-rise, large floor plate building construction common in the United States. The stub girder system involves the use of short sections of beam welded to the top flange of a continuous, heavier bottom girder member. Continuous transverse secondary beams and ducts pass through the openings formed by the beam stubs. This system has been used in many building projects, but generally requires a shored design with consequent construction cost premiums.

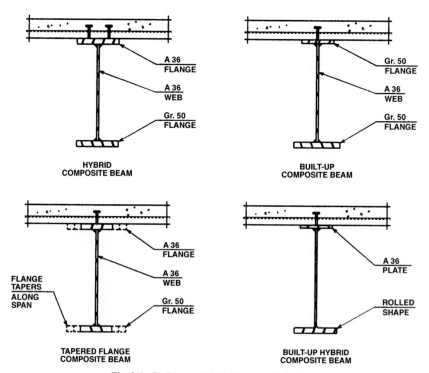

Fig. 3.4   Built-up and hybrid composite beams.

Successful composite beam design requires the consideration of various service-ability issues such as long-term (creep) deflections and floor vibrations. Of particular concern is the issue of perceptibility of occupant-induced floor vibrations. The rela-tively high flexural stiffness of most composite floor framing systems results in rela-tively low vibration amplitudes from transitory heel-drop excitations and therefore is effective in reducing perceptibility. Recent studies have shown that short [7.6 m (25 ft) and less] and very long clear-span [13.7 m (45 ft) and longer] composite floor framing systems perform quite well and are rarely found to transmit annoying vibra-tions to the occupants. Particular care is required for span conditions in the (9.1- to 10.7-m) (30- to 35-ft) range. Anticipated damping provided by partitions which extend to the slab above, services, ceiling construction, and the structure itself are used in conjunction with state-of-the-art prediction models to evaluate the potential for per-ceptible floor vibrations.

2 Composite Joists and Trusses

Preengineered proprietary open-web floor joists, joist girders, and fabricated floor trusses are viable composite members when combined with a concrete floor slab. The advantages of an open-web floor framing system include increased spannability and stiffness due to the deeper structural depth and ease in accommodating electrical con-duit, plumbing pipes, and heating and air-conditioning ductwork. Open web systems do, however, carry a premium for fireproofing the many, relatively thin, components of

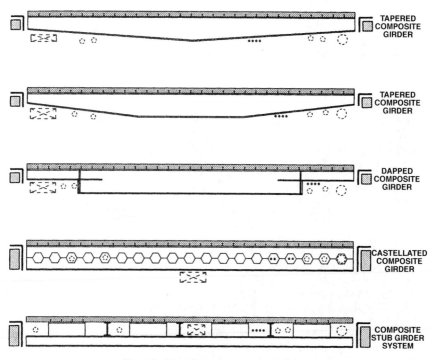

Fig. 3.5 Nonprismatic composite girders.

the member. Open-web steel joists have been used in composite action with flat-soffit concrete slabs and metal deck slabs supporting concrete fill with and without shear connectors. The design for these systems is primarily based on manufacturers' test data. As open-web steel joists and joist girders normally are spaced relatively closely, the full potential for composite efficiency is not realized as compared to other composite floor systems. Composite design does provide quantifiable advantages over non-composite design for open-web floor joists such as increased stiffness and ductility.

Built-up fabricated composite floor trusses combine material efficiency in relatively long-span applications with maximum flexibility for incorporating building-services ductwork and piping into the ceiling cavity. The profile of the truss form allows for large mechanical air ducts as well as other piping and electrical lines to pass through the openings formed by the triangularization of the web members. The increased depth of the composite truss system over a standard rolled-shape composite beam system with building-services ductwork and piping passing below the beam results in maximum material efficiency and high flexural stiffness. Generally, composite floor trusses are considered economically viable for floor spans in excess of about 9 m (30 ft). A further requirement for floor truss systems is that the framing layout be uniform, resulting in relatively few truss types, which can be readily built in the fabrication shop using a jig. Otherwise the high level of fabrication inherent in the floor truss assemblage tends to offset the relative material efficiency. For this reason, composite floor truss systems are particularly attractive in high-rise office building applications where large open lease spans are required and floor configurations are generally repetitive over the height of the building. Figure 3.6 shows an example of a project utilizing composite floor trusses as part of an overall mixed steel and concrete building frame.

Any triangulated open-web form can be used to define the geometry of the fabricated floor truss; however, the Warren truss, with or without web verticals, is the one utilized most often (Fig. 3.7). The Warren truss without verticals provides a maximum open-web area to accommodate ductwork and piping. Vertical web members added to the Warren truss or a Pratt truss geometry may be utilized when the unbraced length of the compression chord is critical. Often a Vierendeel panel in the low-shear zone near the center of the span is incorporated into the truss configuration to accommodate the main air-handling mechanical duct loop in office building applications. The spacing of the web members should be chosen such that the free passage of ductwork and piping is not inhibited while maintaining a reasonable compression top-chord unbraced length. On the other hand, the angle of the web diagonals should be made relatively shallow to reduce the number of members and associated joint welding. This must be balanced by the fact that shallower web members result in longer unbraced lengths and higher member axial forces, often requiring connection gusset plates, thereby increasing fabrication costs and decreasing the clear area for ductwork and piping. A panel spacing of roughly two to three times the truss depth is a good rule of thumb for orienting web diagonals. The floor truss configuration should be detailed such that any significant point loads are applied at truss panel points. A vertical web member may be introduced into the truss girder geometry to transfer these imposed shear loads into the truss system.

A variety of chord and web member cross sections may be utilized in building up the floor truss geometry (see Fig. 3.8). Chord members may be wide-flange T or single-angle sections to allow easy, direct connection of web members without gusset plates. Rectangular tubes or double-angle sections are less commonly used chord members as they require gusset-plated connections. Web members are most often Ts or single- or double-angle sections welded directly to the chord T or angle stem, although tube sections have been used. The composite floor truss system is completed through the direct connection of the top chord flange to the concrete floor slab by shear connectors. The most common floor system in building construction is a composite metal deck and concrete slab chosen based on fire separation and acoustical requirements spanning between composite floor trusses. The floor trusses are normally spaced such that the metal deck slab spans as the concrete form between the trusses without requiring any additional shoring.

3.2 PRESTRESSED AND POSTTENSIONED CONCRETE FLOOR SYSTEMS

Prestressed floors are commonplace in buildings throughout the world, particularly in low-rise structures such as parking garages and shopping centers. Precast pretensioned floor units have remained popular since the 1960s, and cast-in-place posttensioned concrete floors have gained wide acceptance since the mid 1970s.

Posttensioned floors have been widely used for high-rise office buildings in Australia since the early 1980s, and there are examples in the United States, the most notable being 311 South Wacker Drive, Chicago, which was the tallest concrete building in the world when completed.

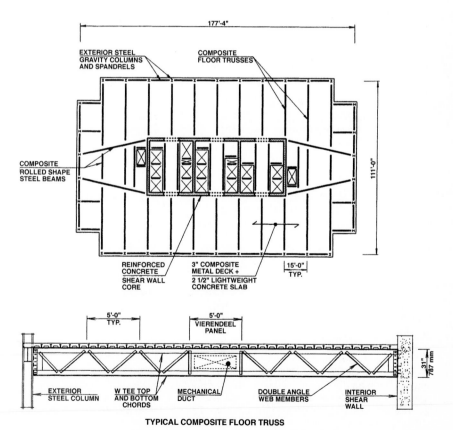

TYPICAL COMPOSITE FLOOR TRUSS

Fig. 3.6   One North Franklin, Chicago, Illinois.

1 General Considerations

High-rise office buildings usually have long-span floors to achieve the desirable column-free space, and the spans are usually noncontinuous between the core and the
facade. To achieve long spans and still maintain acceptable deflections requires a deep
floor system in steel or reinforced concrete. However, by adopting prestressed post-

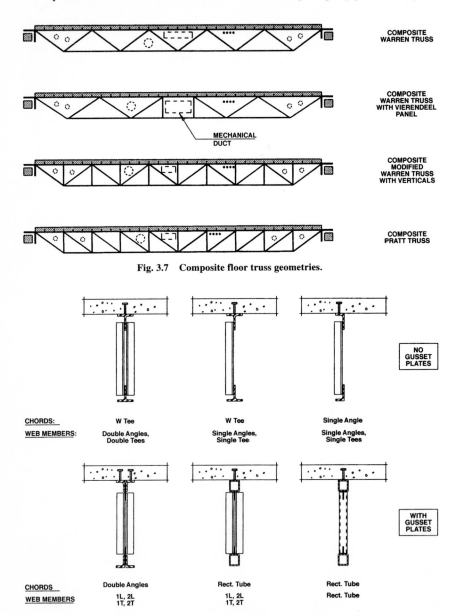

Fig. 3.7   Composite floor truss geometries.

Fig. 3.8   Composite truss component sections.

tensioned concrete beams it is possible to achieve a shallow floor structure and still maintain acceptable deflections without the need for expensive precambering.

High-rise residential buildings usually do not require long spans because column-free space is not a selling point; the tenant or buyer sees the space already subdivided by walls, which effectively hide the columns. Hence continuous spans can be achieved. Unlike office buildings, residential buildings do not as a rule have sus-pended ceilings—the ceiling may be just a sprayed high-build coating on the slab sof-fit or a plasterboard ceiling on battens fixed to the slab soffit. Flat-plate floors are therefore required and deflection control is an important design consideration. Where the columns form a reasonably regular grid, prestressing can be very effective in mini-mizing the slab thickness while at the same time controlling deflections.

Although it is customary to use posttensioning for prestressed concrete high-rise buildings, precast pretensioned concrete can be used and has been employed in some buildings described in this Monograph (Luth Building; Marriott Hotel, New York; Taj Mahal Hotel). The major disadvantage of precast pretensioned concrete floor beams or slabs is the cranage required to lift the heavy units along with the field-welded connec-tions required for stability and diaphragm action. Precast pretensioned floor members are usually tied together by and made composite with a thin cast-in-place topping slab.

Floor posttensioned systems use either 12.7- or 15.2-mm (0.5- or 0.6-in.) high-strength steel strand formed into tendons. The tendons can be either "unbonded," where individual strands are greased and sheathed in plastic, or "bonded," where groups of four or five strands are placed inside flat metal ducts that are filled with cement grout after stressing. On a worldwide basis, bonded systems are preferred in high-rise buildings because they have demonstrated better long-term durability than unbonded systems. Although unbonded systems used today have improved corrosion resistance compared to earlier systems, there is still a large number of older buildings that exhibit corrosion problems in their unbonded tendons. Another reason that bonded posttensioned systems are preferred is that cutting tendons for renovations or demolition is both simpler and safer when the tendons are bonded to the concrete. Nevertheless, care must be exercised as it is by no means unknown for tendons speci-fied to be grouted to have had this vital operation omitted. In this aspect, good quality control is essential. Figure 3.9 illustrates a typical posttensioned floor using unbonded tendons, whereas Figs. 3.10 and 3.11 illustrate the construction of a typical postten-sioned floor using bonded tendons.

The most common posttensioned systems are:

- Posttensioned flat slabs and flat plates (Fig. 3.12)
- Posttensioned beams supporting posttensioned slabs (Fig. 3.13)
- Posttensioned beams supporting reinforced concrete slabs (Fig. 3.14)

Currently with computer programs readily available to carry out cracked section analysis of prestressed concrete, it is normal to design for partial prestress where the concrete is assumed to be cracked at full design working load and untensioned steel comprises a significant portion of the total reinforcement. The partial prestress ratio (PPR) gives the degree of prestress

$$\text{PPR} = \frac{A_p f_{py}}{A_p f_{py} + A_s f_{sy}} \tag{3.1}$$

where $A_p f_{py}$ is the cross section area of prestressed steel multiplied by its yield strength and $A_s f_{sy}$ is the cross section area of normal reinforced steel multiplied by its yield stress. A useful starting point in determining the amount of prestress required is to pro-vide sufficient prestress to balance about 75% of the self-weight of the floor structure.

Untensioned steel is then added to satisfy the ultimate limit state. (This will often result in a PPR of about 0.6.) Deflections and shear capacity must also be checked.

The span-to-depth ratio of a single-span noncontinuous floor beam will be about 25; for a continuous beam it will be about 28 and for a flat-plate beam about 45 for an internal span and 40 for an end span.

Fig. 3.9   Typical posttensioned floor using unbonded tendons.

Fig. 3.10   Typical posttensioned floor using bonded tendons.

In high-rise buildings it is preferable to avoid running floor beams into heavily reinforced perimeter columns for two reasons:

1. There are difficulties in accommodating tendon anchorages, which compete for space with the column reinforcement.
2. Frame action developed between the beams and columns causes the design bending moment between floors to vary as the frames resist lateral load, thereby diminishing the number of identical floors that can be designed, detailed, and constructed.

Instead of being directly supported by columns, the floor beams should be supported by the spandrel beams.

Prestressing anchorages can be on the outside of the building (requiring external access), at a step in the soffit of the beams [see Riverside Centre and Bourke Place (Figs. 3.15, 3.30, and 3.33)], or in a pocket at the top of the floor. Top-of-floor pockets have the disadvantage that they usually cause local variations in the flatness of the floor and rough patches, which may need to be ground flush.

Because posttensioning causes axial shortening of the prestressed member, it is necessary to consider the effects of axial restraint, that is, the effects of stiff columns

TENDON PLACING

CONCRETING

STRESSING

GROUTING

Fig. 3.11   Construction sequence for bonded posttensioned concrete.

and walls. Such restraint has two potential effects: it can overstress the columns or walls in bending and shear, and it can reduce the amount of prestress in the floor.

Fortunately the stiff core of a high-rise building is usually fairly central so that the axial shortening of the floor can be generally in a direction toward the core. This means that the perimeter columns move inward, but because they move by the same amount from story to story, no significant permanent bending stresses occur except in the first story above a nonprestressed floor, which is often the ground floor. As this story is usually higher than a typical story, the flexibility of the columns is greater and the induced bending moments may be easily accommodated. However, the loss of pre-stress in the floor may necessitate some additional untensioned reinforcement.

2 Economics of Posttensioning

Posttensioned concrete floors will usually result in economics in the total construction cost because of the following:

- Less concrete used because of shallower floor structure (Fig. 3.16)
- Less load on columns and footings
- Shallower structural depth, resulting in reduced story height (Fig. 3.17)

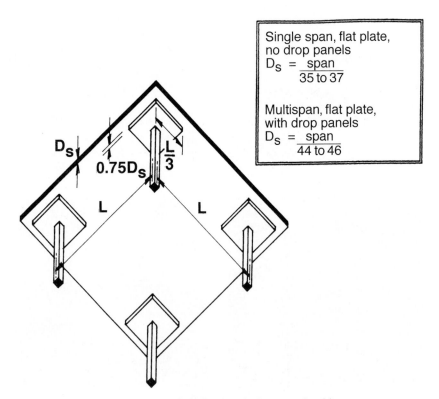

Single span, flat plate, no drop panels
$$D_S = \frac{\text{span}}{35 \text{ to } 37}$$

Multispan, flat plate, with drop panels
$$D_S = \frac{\text{span}}{44 \text{ to } 46}$$

Fig. 3.12 Posttensioned flat slab, single span and multispan.

The last item can be very significant as any height reduction translates directly into savings in all vertical structural, architectural, and building-services elements.

The construction will proceed with the same speed as a normal reinforced concrete floor, with four-day floor-to-floor construction cycles being achieved regularly on high-rise office buildings with posttensioned floors (Fig. 3.18). Three-day cycles can easily be achieved using an additional set of forms and higher strength concretes to shorten posttensioning time.

A major cost variable in posttensioned floors is the length of the tendons. Short tendons are relatively expensive compared to long tendons. Figure 3.19 shows the cost trend for tendons ranging from 10 to 60 m (33 to 200 ft). The relatively high cost of short tendons results from fixed-cost components such as setup costs, anchorages, and tendon stressing being prorated over lesser amounts of strand. The influence of strand "setting losses" is also greater with very short strands, thus increasing the area of tendon required. Nevertheless, even though most tendons in a high-rise building floor will be only around 10 to 15 m (33 to 50 ft), the system is economical because of savings in floor depth, and it is desirable because of control of deflections and the lack of need for precambering. For grouted tendons, the optimum economical size has been found to be the four- or five-strand tendon in a flat duct because the anchorages are compact and readily accommodated within normal building members and because stressing is carried out with a light jack easily handled by one person.

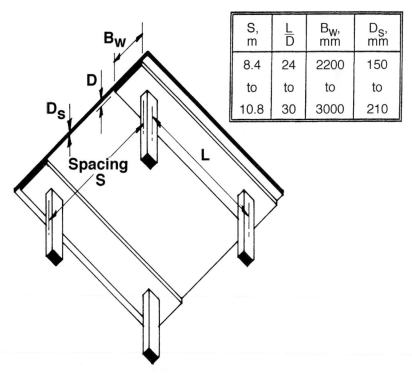

S, m	$\frac{L}{D}$	B$_w$, mm	D$_s$, mm
8.4	24	2200	150
to	to	to	to
10.8	30	3000	210

Fig. 3.13   Posttensioned beam and slab, multispan.

Comparing the cost of bonded and unbonded tendons will generally show the unbonded system as being slightly cheaper. This is because unbonded posttensioning usually requires less strand due to lower friction and greater available drape. Unbonded strand also does not need grouting with its costs of time and labor. As a floor using unbonded strand will require more reinforcement than a bonded system due to lower ultimate flexural strength and code requirements, the combined cost of the strand and untensioned reinforcement will be almost the same as that for bonded systems.

The cost of a posttensioned system is further affected by the building floor geometry and irregularities. For example:

- The higher the perimeter-to-area ratio, the higher the normal reinforcement content since reinforcement in the perimeter can be a significant percentage of the total.

- Angled perimeters increase reinforcement and make anchorage pockets larger and more difficult to form.

- Internal stressing from the floor surface increases costs due to the provision of the wedge-shaped stressing pockets and increased amounts of reinforcement.

- Slab steps and penetrations will increase posttensioning costs if they decrease the length of tendons.

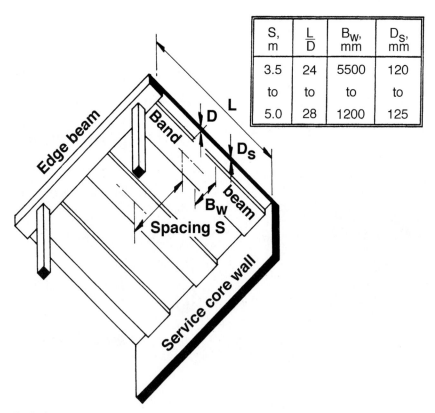

S, m	$\dfrac{L}{D}$	B_W, mm	D_S, mm
3.5 to 5.0	24 to 28	5500 to 1200	120 to 125

Fig. 3.14   Posttensioned beam at close centers with posttensioned or reinforced slab, single span.

Fig. 3.15   Bourke Place, Melbourne, Australia; 53 levels.

3 Cutting Prestressed Tendons

One of the main drawbacks of posttensioned systems is the difficulty of dealing with stressed strands and tendons during structure modifications or demolition. Although modifications are more difficult, some procedures have been developed to make this process easier.

Small penetrations required to meet changes to plumbing or similar requirements are the most common of all modifications that are made to the floor system. The size of these penetrations is typically from 50 to 250 mm (2 to 10 in.) in diameter. As a posttensioned floor relies on the posttensioned tendons for its strength, it is preferable to avoid cutting the tendons when drilling through the floor for the new penetration. Finding the tendons in a floor to permit the location of penetrations without damaging any tendons is a very simple procedure that is carried out with the aid of an electronic tendon locater. Tendons are accurately located using this system without any need to remove floor coverings or ceilings.

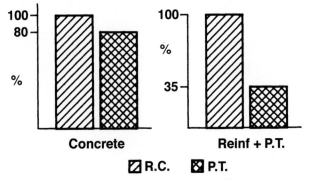

Fig. 3.16   Material handling—reinforced concrete versus posttensioned system.

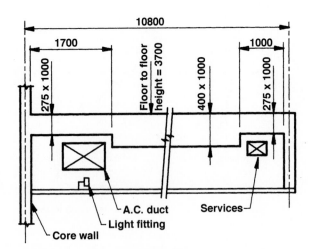

Fig. 3.17   Example of stepped beam soffit; Bourke Place, Melbourne, Australia.

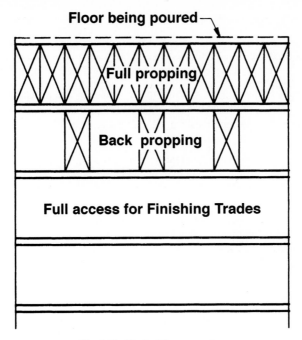

Fig. 3.18   Typical floor propping.

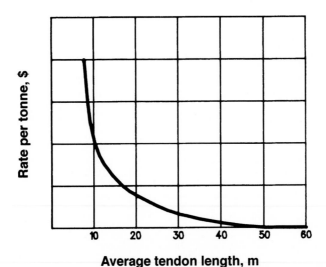

Fig. 3.19   Posttensioning costs.

In a typical posttensioned floor it is possible to locate penetrations of up to 1000 by 3000 mm (3 by 9 ft) between posttensioned tendons and to require no other modification to the floor. Penetrations that require cutting of the posttensioned tendons will need to be checked and designed as would any large penetration in any floor system. The procedure commonly adopted in a floor using bonded tendons is as follows:

1. Design the modified floor structure in the vicinity of the penetration, assuming that any cut posttensioned tendons are dead-ended at the penetration.
2. Install any strengthening required.
3. Locate tendons and inspect grouting.
4. If there is no doubt as to the quality of the grouting, proceed to step 5. Otherwise strip off ducting, clean out grout, and epoxy grout the strands over a length of 500 mm (20 in.) immediately adjacent to the penetration.
5. Install props.
6. Core drill the corners of the penetration to eliminate the need for overcutting, and then cut the perimeter using a diamond saw.
7. Cut up the slab and remove.
8. Paint an epoxy-protective coating over the ends of the strands to prevent corrosion.
9. Remove props.

If a large penetration through a floor cannot be located within the slab area but must intersect a primary support beam, then substantial strengthening of adjacent beams will usually be necessary.

When cutting openings into floors built using unbonded posttensioned tendons the procedures used for bonded posttensioned tendons cannot be applied. The preferred procedure that has been developed to permit controlled cutting of unbonded strands is to use a special detensioning jack. The jack grips the strand and the strand is then cut, with the force in the strands being released slowly. New anchorages are then installed at each side of the new opening and the strands restressed.

Extensive experience has been gained in demolition procedures for posttensioned floors, and some general comments can be made. In bonded systems the procedures for demolition are the same as for reinforced concrete. The individual strands will not dislodge at stressing anchorages. In unbonded systems the strand capacity is lost over its entire length when cut; therefore the floor will require backpropping during demolition. The individual cut strands will dislodge at stressing anchorages, but will move generally less than 450 mm (18 in.). However, precautions should always be taken in case the strands move more than this.

PROJECT DESCRIPTIONS

Melbourne Central
Melbourne, Australia

Architect	Kisho Kurokawa with Bates Smart & McCutcheon
Structural engineer	Connell Wagner
Year of completion	1991
Height from street to roof	211 m (692 ft)
Number of stories	54
Number of levels below ground	3
Building use	Office
Frame material	Concrete core, steel floor beams
Typical floor live load	3-kPa (60-psf) beams, 4-kPa (80-psf) slabs
Basic wind velocity	50 m/s (112 mph) ultimate, 100-yr return
Maximum lateral deflection	100 mm (4 in.), 50-yr return
Design fundamental period	4.2 sec
Design acceleration	2.9 mg rms, 5-yr return
Design damping	1% serviceability, 5% ultimate
Earthquake loading	Not applicable
Type of structure	Concrete core, concrete perimeter tube in tube
Foundation conditions	Mudstone, 2000-kPa (20-ton/ft^2) capacity
Footing type	Pads to columns, raft to core
Typical floor	
Story height	3.85 m (12 ft 7 in.)
Beam span	11.5 m (37 ft 9 in.)
Beam depth	530 mm (21 in.)
Beam spacing	3 m (10 ft)
Slab	120 mm (4.75 in.) on metal deck
Columns	
Size at ground floor	1000 by 1200 mm (39 by 47 in.)
Spacing	6 m (20 ft)
Concrete strength	60 MPa (8500 psi) maximum
Core	
Shear walls	65 MPa (10,000 psi) maximum
Thickness at ground floor	600 and 200 mm (24 and 8 in.)

Melbourne Central comprises a 57-level office tower of 60,000 m^2 (646,000 ft^2) (net rentable) and a large retail development of a further 60,000 m^2 (Fig. 3.20). The overall dimensions of the tower are 43.72 by 43.72 m (143 by 143 ft). The tower is 211 m (692 ft) above street level and 225 m (738 ft) above the core raft. The facade is a glass and aluminum curtain wall.

The tower floors consist of steel beams spanning from the core to the facade with a composite concrete slab, supported on structural steel decking, spanning between the steel beams (Fig. 3.21). The steel beams are generally at 3-m (10-ft) centers, and the typical beam is a 530UB82 (21UB55). The structural steel decking is 1 mm (0.04 in.) thick, unpropped.

Fig. 3.20   Melbourne Central, Melbourne, Australia.

The column spacing at the facade is 6 m (20 ft). A perimeter beam is required to carry the intermediate floor beams. This is a 900-mm-deep by 300-mm-wide (36- by 12-in.) precast concrete beam. Although this is precast concrete, it is erected in the same way as a steel beam and as part of the steel frame. The use of precast concrete simplifies the fire rating of the structure at the perimeter where access is difficult. It also provides the 900-mm (36-in.)-deep fire barrier between floors required by the building regulations. The fixings for the curtain wall are cast into this beam, resulting in reliable and accurate positioning.

The floor-to-floor height is 3875 mm (12 ft 8.5 in.) for the typical floors. The floor-to-ceiling height is 2900 mm (9 ft 6 in.), which allows for a future access floor of 200 mm (8 in.) in height, to be installed by a tenant, providing a minimum 2700-mm (8-ft 10-in.) occupied space.

The wind resistance structure for this building consists of the core cantilevering from the footing in combination with a nominal contribution from the facade structure of the column and precast beam. This results in the facade structure carrying approximately 10% of the wind load on the building, and, more importantly, it contributes

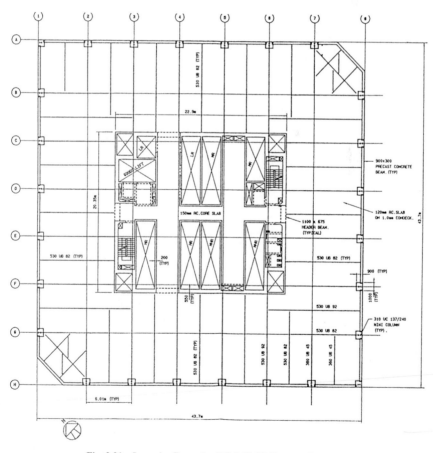

Fig. 3.21   Low-rise floor plan L8–L14; Melbourne Central.

significantly to the sway serviceability performance. The remainder of the wind load is carried by the core element.

The central-services core to the building is reinforced concrete from the footings to the roof. All the internal walls are 200 mm (8 in.) thick. This thickness remains unchanged over the full height of the building. The 200-mm (8-in.) internal wall thickness is the optimum to achieve load-carrying capacity, minimal slenderness effects, and constructability. The external walls vary from 600 and 550 mm (24 and 22 in.) thick at the bottom of the building to 250 mm (10 in.) thick at the building top. Concrete strengths in the core walls vary from 70 to 30 MPa (10,000 to 4300 psi) at 90 days.

The columns are a composite of reinforced concrete with a 310UC137 steel column. These steel columns are erected as part of the steel frame. Subsequently they are encased within the reinforced concrete column and permit erection of the steel frame 10 floors ahead of concrete encasement (Fig. 3.22). This concept, in combination with the steel floor beams and structural steel decking, permits benefiting from the advantages of steel construction while at the same time minimizing the quantity of the relatively expensive material that is steel. This is fundamental to a composite steel and concrete building of this type, where the advantages of reinforced concrete and steel are both incorporated into the structure.

The footings to the tower are founded in moderately weathered mudstone having a bearing capacity of 2000 kPa (20 ton/ft^2). The depth of the excavation and the basement is such that the footings at the west end of the tower are founded near the top of this material. The footing to the core is a 3.2-m (10-ft 6-in.)-thick reinforced concrete raft. This extends approximately 2 m (6 ft 6 in.) past the outside face of the core wall.

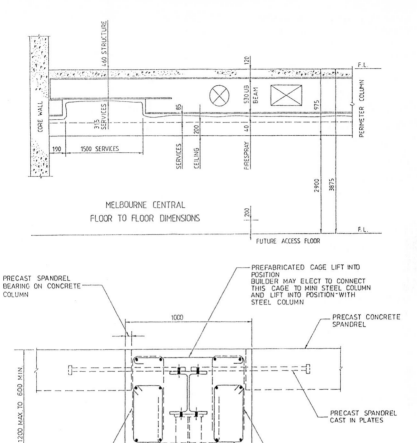

MELBOURNE CENTRAL
FLOOR TO FLOOR DIMENSIONS

Fig. 3.22   Floor-to-floor dimensions and typical tower column reinforcement details; Melbourne Central.

Luth Headquarters Building
Kuala Lumpur, Malaysia

Architect	Hijjas Kasturi Associates
Structural engineer	Ranhill Bersekutu
Year of completion	1984
Height from street to roof	152 m (498 ft)
Number of stories	38
Number of levels below ground	0
Building use	Offices, parking garage
Frame material	Concrete
Typical floor live load	2.5 kPa (50 psf)
Basic wind velocity	30 m/s (67 mph)
Maximum lateral deflection	Not available
Design fundamental period	Not established
Design acceleration	Not established
Design damping	Not established
Earthquake loading	Not applicable
Type of structure	Tube in tube
Foundation conditions	Stiff silty clay
Footing type	1500-mm (5-ft)-diameter bored piles, 20 m (60 ft) deep

Typical floor
Story height	3.66 m (12 ft)
Beam span	Varies from 19.2 to 8.7 m (64 to 28.5 ft)
Beam depth	Typically 640 mm (25 in.)
Beam spacing	9 degrees radially
Material	Precast pretensioned concrete
Slab	100 mm (4-in.) precast planks, 50-mm (2-in.) topping

Columns
Size at ground floor	5 by 1.2 m (16.4 by 4 ft)
Spacing	38 m (125 ft) around circumference
Concrete strength	32 MPa (5000 psi)
Core	Reinforced slip-formed concrete
Thickness at ground floor	400 and 200 mm (16 and 8 in.)
Concrete strength	32 MPa (5000 psi)

The Luth Headquarters Building is a 38-level office building in Kuala Lumpur (Fig. 3.23). Of the 38 levels, 37 are at or above ground and comprise 7 levels of parking garage, 2 mechanical-plant levels, and 28 levels of office space.

All floors are circular and contain a circular central core. However, in elevation the building is most unusual in that the facade is not vertical but formed from several solids of revolution. The facade of the lowest 22 levels is described by one circular

Fig. 3.23 Luth Headquarters Building, Kuala Lumpur, Malaysia.

curve, creating progressively smaller floors with clear spans from core to facade vary-ing from 19.2 m (63 ft) at the first floor to 8.7 m (28 ft 6 in.) at the nineteenth floor. The facade from level 19 to 22 is described by a second circular curve which provides a transition into the inverted conical shape of the upper 15 levels, where floor spans increase to a maximum of 12.2 m (40 ft).

The building is supported by a central core and 40 perimeter columns. These columns are supported on a massive arched transfer structure at the first-floor level with only five points of support. Each of the five arch supports has a design vertical load of 8000 tonnes (8800 tons) and a horizontal load of 2000 tonnes (2200 tons). The vertical load is resisted by ten 1500-mm (5-ft)-diameter piles drilled some 20 m (60 ft) into the clay. The horizontal load is resisted by prestressed concrete tie beams which interconnect all five footings.

The foundation conditions comprise silty clay for the full depth of 60 m (197 ft) investigated, with the clay strength decreasing slightly with depth. Of the spread foot-ings, a raft, and piles considered, piles were chosen as being the most suitable for the structure.

A report by a specialist geotechnical consultant predicted a possible 50-mm (2-in.) differential settlement between core and perimeter columns, so the tie beams were arranged in a manner to bypass the core, resulting in the star shape seen in Fig. 3.24. The beams, which are 900 by 900 mm (36 by 36 in.) and 55 m (180 ft) long, were stressed in three stages to approximately balance the ever-increasing horizontal thrusts as the construction progressed. Prestressing was carried out from within large man-holes provided for that purpose.

The settlement at the time of completion of the construction was nowhere in excess of 3 mm ($^1/_8$ in.), which was also the maximum differential, so the 50-mm (2-in.) value will most likely never occur. Nevertheless, permanent settlement monitoring plates have been fixed to columns and the core to facilitate future periodic survey.

The reduction from 40 columns in the upper level to only five below the first floor is achieved by two major transfer structures: a 3.8-m-deep by 0.9-m-wide (12 ft 6 in. by 3 ft) prestressed concrete belt around the eighth-floor plant room and five continu-ous arches between the ground floor and the first floor. The arches vary in depth from 9 to 1.8 m (30 to 6 ft) and are a constant 1.2 m (4 ft) thick.

The 1.8-m (6-ft) depth at the crown was based on architectural limitations and proved grossly inadequate to support the load of the upper 35 levels. The eighth-floor ring beam was therefore created to bridge across the flexible central part of the arches.

The ring beam is subject to particularly high moments and shears. A partially pre-stressed design proved to be ideal, as it was possible to anchor tendons at the beam soffit such that their steep rise to the top of the beam in the negative-moment region provided a significant contribution to shear resistance. A total of ten 27-strand ten-dons, each of exactly the same length, was used in an arrangement that provided four tendons in negative-moment regions and two for positive moments. To ensure that the prestress was applied to the ring beam and was not absorbed by other elements, the ninth floor was temporarily separated from the beam and supported on sliding bear-ings, and the columns were provided with temporary concrete hinges.

All floors above the first are totally precast (except for the facade); 40 radial beams support precast concrete slabs, which are topped in situ to create a composite system. Beams are unequal I's partially prestressed, and except for some of the longest spans, are erected unpropped. Their overall depth is 610 mm (24 in.) for park-ing-garage floors and 483 mm (19 in.) for office floors. Figure 3.25 shows a typical mid-rise floor plan.

Because of the predicted differential settlement between core and facade, the floor beams were supported on steel brackets at each end, allowing them to hinge without undue distress. Figure 3.26 shows the typical floor section, and Fig. 3.27 shows the core to floor beam joint. Figure 3.28 gives the building section.

The outer end of the beams resists the radial component of the force from the sloping columns, which varies from a compression of 388 kN (42 tons) to a tension of 109 kN (12 tons) per beam. The force is ultimately resisted by the floor acting as a stiff, flat ring.

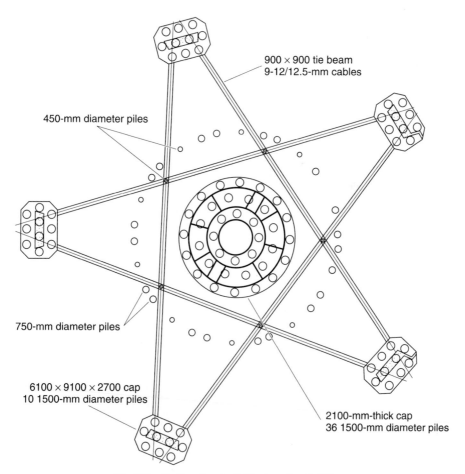

900 × 900 tie beam
9-12/12.5-mm cables

450-mm diameter piles

750-mm diameter piles

6100 × 9100 × 2700 cap
10 1500-mm diameter piles

2100-mm-thick cap
36 1500-mm diameter piles

Fig. 3.24   Footing plan; Luth Headquarters Building.

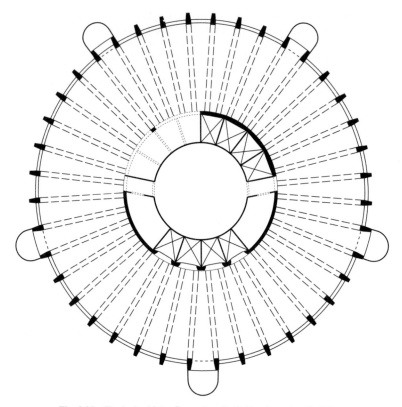

Fig. 3.25   Typical midrise floor plan; Luth Headquarters Building.

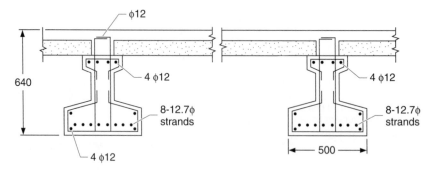

Fig. 3.26   Typical floor section; Luth Headquarters Building.

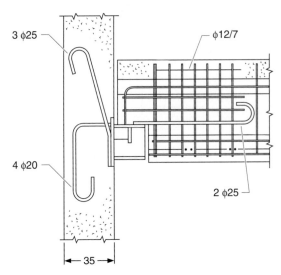

Fig. 3.27   Core to floor beam joint; Luth Headquarters Building.

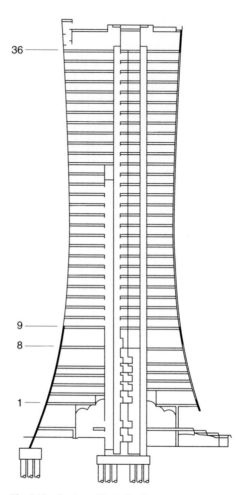

Fig. 3.28   Section of Luth Headquarters Building.

Riverside Center
Brisbane, Australia

Architect	Harry Seidler & Associates
Structural engineer	Rankine & Hill
Year of completion	1986
Height from street to roof	150 m (492 ft)
Number of stories	39
Number of levels below ground	2
Building use	Office
Frame material	Concrete
Typical floor live load	4 kPa (80 psf)
Basic wind velocity	50 m/s (112 mph)
Maximum lateral deflection	63 mm (2.5 in.), 50-yr return
Design fundamental period	3.8 sec
Design damping	2% serviceability, 5% ultimate
Earthquake loading	Not applicable
Type of structure	Tube in tube
Foundation conditions	Rock, 5-MPa (56-ton/ft^2) capacity
Footing type	Pads to columns, mat to core
Typical floor	
Story height	3.475 m (11 ft 5 in.)
Beam span	12 m (39 ft 4 in.)
Beam depth	600 mm (24 in.)
Beam spacing	3.35 m (11 ft)
Material	Posttensioned concrete
Slab	125 mm (5 in.) reinforced concrete
Columns	
Size at ground floor	1100 by 700 mm (43 by 27 in.)
Spacing	6.7 m (22 ft)
Concrete strength	50 to 32 MPa (7200 to 4500 psi)
Core	Concrete shear walls
Thickness at ground floor	350 and 200 mm (14 and 8 in.)
Concrete strength	40 to 25 MPa (5700 to 3500 psi)

This 39-story, 42-level building is a totally reinforced concrete structure designed as a "tube in tube" (Fig. 3.29). However, because the triangular shape leads to unusually long exterior core walls, the core has a greater than normal stiffness, and the exterior spandrel beams and columns play only a minor role in the resistance to wind load (Fig. 3.30). The floors are supported by simply supported partially prestressed beams spanning 12 m (40 ft) from core to perimeter. Slabs are not prestressed.

Apart from the office building, the development includes a two-level basement garage, which covers the site and extends into the Brisbane River. The lowest floor is below normal high-tide levels, and the whole basement is designed to continue to function normally during a flood of a height resulting in a head of 6 m (20 ft) of water at the lowest floor. The garage is topped by a ground-level plaza, low-rise commercial and retail buildings, and a restaurant which cantilevers 14 m (46 ft) over the river.

Fig. 3.29   Riverside Center, Brisbane, Australia.

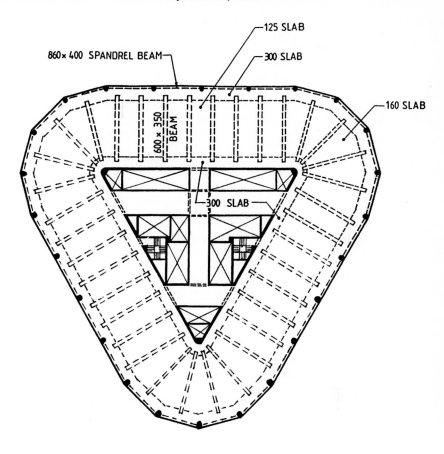

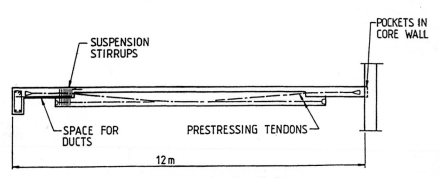

Fig. 3.30   Floor plan; Riverside Center.

The ground conditions comprise hard phyllite, a metamorphosed mudstone, which allowed the use of design bearing pressures of 5000 kPa (50 ton/ft^2). Footings for the tower are reinforced concrete pads to columns and a raft slab to the core. The surrounding basement columns are supported on either pads or piers, depending on the rock level, which sloped away into the river.

Floor slabs are designed for a general live load of 4 kPa (80 psf) with a 5-kPa (100-psf) zone around the perimeter of the core. The use of 4 kPa (80 psf) rather than the statutory 3 kPa (60 psf) provides for the more ready accommodation of safes, isolated compacting units, and other heavy loads over a small area. The 125-mm (5-in.)-thick slabs span 3.3 m (10.8 ft) and are reinforced with fabric.

Floor beams are 600 mm (24 in.) deep and 350 mm (14 in.) wide at the soffit. (Sides are tapered to ease form removal.) At each end the beams terminate in a 300-mm (12-in.)-thick slab, leaving about a 1200-mm (4-ft)-wide zone in which to locate major air distribution ducts. The prestressing tendons, of which there are two per beam, usually four-strand, are contained in circular ducts, but anchored in slab type anchorages. The slab anchorages are the most economical and lend themselves to the use of small, light jacks. The circular ducts result in narrower beams compared with the width required for two flat slab ducts side by side.

The partially prestressed design provides for a load-balanced condition for about 80% of the weight of the bare concrete. This resulted in a flat floor. Ultimate load capacity was provided by additional untensioned steel. Untensioned steel stresses were limited to 150 MPa (21,400 psi). Beams were designed for the same live loads as the slabs, except that reductions in accordance with the loading code were used.

At each end of the beam, where it becomes a wide 300-mm (12-in.)-deep slab, considerable analysis effort was undertaken to ensure satisfactory stress levels. Here reinforcement is predominantly untensioned steel, with only one of the tendons continuing to the supporting spandrel beam; the other tendon terminates in a stressing anchorage at the end of the 600-mm (24-in.)-deep section of the beam. This arrangement of tendons provided for stressing off the floor below—there were no external scaffolding requirements.

Stressing was carried out in two stages: 50% 3 days after pouring the slab and 100% after 7 days. These requirements dictated the concrete strength rather than the minimum design strength specified. [The concrete yielded a strength of about 35 MPa (5000 psi) at 28 days, with 25 MPa (3500 psi) having been specified.] A prop load analysis was carried out, taking into account the load-relieving effect of the prestress, in order to arrive at the time when props could be removed.

Plant-room beams support a much heavier load than office floor beams, but the same floor formwork could be utilized by increasing the slab thickness and overall beam depth and by sloping the floor surface upward from the midspan of the beams. (The slab had to be thicker for acoustic reasons anyway, and a fall for drainage was always required, so the structural requirements matched the other requirements.)

The service core has concrete walls, generally 200 mm (8 in.) thick, except for the perimeter walls, which vary from 350 to 300 to 250 mm (14 in. to 12 to 10 in.). Some tension in the lower stories occurs under design wind loads, but in general loads are compression. Concrete was pumped for the full 150-m (492-ft) height, with strengths varying from 40 to 25 MPa (5700 to 3500 psi).

An architectural limit was placed on the column sizes, resulting in the use of 50-MPa (7100-psi) concrete and 4% reinforcement at the lower levels. Some early problems were encountered with misplaced bars, which made the placing of spandrel beam reinforcement very difficult, particularly as the column bars were 36 mm (1.4 in.) in diameter and in bundles of up to four bars, but once a steel template was employed to locate the bars, the problems disappeared. Where bundled bars were used, all column bars were specified to have splicing sleeves.

Bourke Place
Melbourne, Australia

Architect	Godfrey & Spowers
Structural engineer	Connell Wagner
Year of completion	1991
Height from street to roof	223 m (732 ft)
Number of stories	54
Number of levels below ground	3
Building use	Office
Frame material	Concrete
Typical floor live load	4 kPa (80 psf)
Basic wind velocity	39 m/s (87 mph), 50-yr return
Maximum lateral deflection	200 mm (8 in.), 50-yr return
Design fundamental period	4.8 sec
Design acceleration	3.7 mg rms, 5-yr return
Design damping	1% serviceability, 5% ultimate
Earthquake loading	Not applicable
Type of structure	Reinforced concrete core and perimeter frame tube-in-tube
Foundation conditions	Highly weathered siltstone
Footing type	Pads to columns, raft to core
Typical floor	
Story height	3.7 m (12 ft 2 in.)
Beam span	10.8 m (35 ft 5 in.)
Beam depth	400 mm (16 in.)
Beam spacing	4.6 m (15 ft)
Material	Posttensioned concrete
Slabs	125-mm (5-in.) reinforced concrete
Columns	
Size at ground floor	1100 mm (43 in.) square
Spacing	8.1 m (26 ft 6 in.)
Concrete strength	60 MPa (8500 psi) maximum
Core	Slip-formed shear walls
Thickness at ground floor	400 and 200 mm (16 and 8 in.)
Concrete strength	60 MPa (8500 psi) maximum

The Bourke Place project includes a tower structure with 54 floors above Bourke Street in the city of Melbourne (Fig. 3.31). On top of the concrete tower is a steel-framed, aluminum-clad cone roof reaching another four stories and a communications tower rising to approximately 255 m (837 ft) above the street. Alongside the tower there are an 8-story parking garage (four of which are below ground) and plazas with food and retail areas. The total leasable floor space in the office tower is approximately 60,500 m² (651,200 ft²).

The tower structure consists of a slip-formed reinforced concrete core, posttensioned concrete band beams, and a reinforced concrete perimeter frame (Figs. 3.32

and 3.33). The core structure is approximately 20 m (66 ft) square at the base. Most internal walls are 200 mm (8 in.) thick, with some 150 mm (6 in.), and remain constant for the full height of the structure. The external walls vary from primarily 400 mm (16 in.) thick at the base, using 60-MPa (8500-psi) concrete, to 200 mm (8 in.) for the top 15 stories, requiring only 25-MPa (3500-psi) concrete [40 MPa (5500 psi) was used for pumpability.]

The use of high-strength 60-MPa (8500-psi) concrete allowed the wall thicknesses to be minimized. It was estimated that the loss of floor space for thicker walls, if 40-

Fig. 3.31   Bourke Place, Melbourne, Australia. (*Photo by Squire Photographics.*)

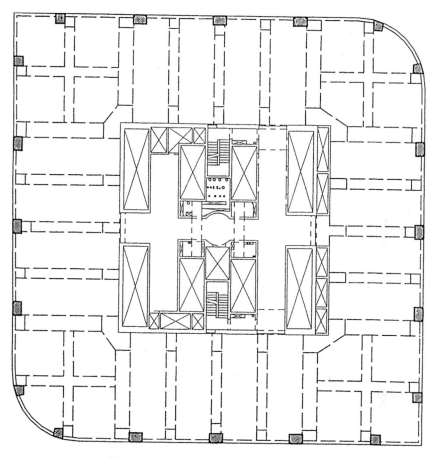

Fig. 3.32   Typical tower floor plan; Bourke Place.

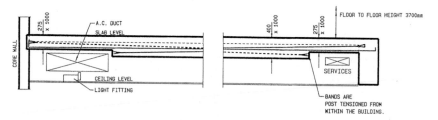

Fig. 3.33   Typical floor profile, Bourke Place.

MPa (5500-psi) concrete was used, represented an effective extra overall capitalized cost to the client of approximately $100,000 (Australian) per floor.

Two substantial core shape changes occur up in the tower as elevator shafts that service the lower levels become redundant. The location of these shape changes and the changes in wall thickness were positioned sufficiently high up in the tower to ensure that the core would be off the construction critical path in order to avoid any time delays. The design of the slip form incorporated the facility to reduce the wall thickness and to "drop off" these portions. Cost comparisons during the design development phase indicated that slip forming was the most cost-efficient method of construction, and the Bourke Place core was the largest single slip-formed core ever constructed in Australia. The core construction set an Australian record in November 1989 for pumping concrete to a total rise of 236 m (774 ft).

At the time of design, building regulations for fire protection required that spandrel beams be at least 900 mm (36 in.) deep. It was recognized that, in conjunction with the columns, these beams would therefore make some contribution to the overall resistance to wind loads on the structure. The beams were designed for the dead and live load requirements; then their capacity to resist additional wind load was assessed. This amounted to approximately 7.5% of the total wind load on the structure, meaning that the core need only be designed for 92.5% rather than the full wind load. The "core and partial-frame" approach represented significant cost savings to the client.

A 125-mm (5-in.) normally reinforced concrete slab spans between 10.8-m (35.4-ft)-long band beams at typically 4.6-m (15-ft) centers. The band beams radiate out from the core and are typically 400 mm (16 in.) deep, but are notched at each end to 275 mm (11 in.). The notches were introduced to accommodate primary mechanical ducts, and they enable the total floor-to-floor height to be minimized. This represents savings to the client as the overall height of the building can be reduced without affecting the number of floors.

The band beams are posttensioned from underneath, utilizing the vertical face of the notches. This separates the posttensioning contractors from the "work face," allowing stressing to be carried out independent of scaffold erection on the newly poured floor, and it eliminates the need for recessed pockets in the floor surface.

The builder used three sets of table forms which "leapfrogged" up the structure and divided the floor into four pours of approximately 350 m² (3800 ft²), with the intention of pouring one quadrant every day. To assist in maintaining this 4-day cycle, column and beam reinforcement cages were standardized where possible and prefabricated.

The floors were checked to ensure that under the most favorable circumstances no back propping would be necessary. Typically, floor cycles of approximately 4 to 5 working days were achieved.

**Central Plaza One
Brisbane, Australia**

Architect	Dr. Kisho Kurokawa, Peddle Thorp Partnerships
Structural engineer	Maunsell Pty. Ltd.
Year of completion	1988
Height from street to roof	174 m (571 ft)
Number of stories	44
Number of levels below ground	4
Building use	Office
Frame material	Concrete
Typical floor live load	3 kPa (60 psf)
Basic wind velocity	49 m/s (110 mph), 50-yr return
Maximum lateral deflection	350 mm (13.75 in.), 25-yr return
Design fundamental period	4.4 and 3.8 sec
Design acceleration	16 mg peak, 5-yr return
Earthquake loading	Not applicable
Type of structure	Central core with perimeter framed tube
Foundation conditions	Marine clay over rock, 5-MPa (5-ton/ft^2) capacity
Footing type	Spread footings, anchored perimeter wall
Typical floor	
Story height	3.66 m (12 ft)
Slab	10-m (33-ft)-span posttensioned, 275 mm (10.8 in.) thick
Columns	
Size at ground floor	1200 by 1000 mm (47 by 39 in.)
Spacing	7 m (23 ft)
Concrete strength	50 MPa (7100 psi)
Core	Concrete shear walls
Thickness at ground floor	600 and 250 mm (24 and 10 in.)
Concrete strength	50 to 32 MPa (7100 to 4600 psi)

Central Plaza One is currently Brisbane's tallest building with a total of 48 levels and has a total height of approximately 174 m (571 ft) above street level (Fig. 3.34).

The building features a four-story atrium with an internal running stream and landscaping at the ground-floor level, and a four-level basement garage. A distinctive roof line with a lifting, slewing telescopic building maintenance unit forming the top 2.5 m (8 ft) of the roof structure makes the building unique among modern high-rise buildings in Australia. The tower houses three plant rooms at levels 4, 26, and 41.

A six-story office block adjacent to the main tower has banking facilities at the ground-floor level and shares the common basement structure with the tower. This "bank annex" incorporates an additional plant room at level 5.

The tower structure comprises a reinforced concrete core and frame with posttensioned floors and is founded on rock approximately 13 m (43 ft) below street level. Design requirements were as follows:

- Column-free office space requiring floors to span 10 m (33 ft) from perimeter beams to central core
- Floors to be designed to allow for maximum flexibility in locating penetrations for services
- Floor edge beams to be designed and detailed to allow for variations at corners to range from 6-m (20-ft) cantilevers to fully truncated corners
- A minimum number of minimum-size columns up through the atrium and above together with the assurance that accelerations due to wind-excited oscillations be within acceptable human response limitations

Fig. 3.34   Central Plaza One, Brisbane, Australia.

- An accurate assessment of deformations due to creep, shrinkage, and load effects to allow for joint design at critical locations in the curtain-wall system
- A basement structure to accommodate 270 cars
- A roof structure to support a lifting, slewing, and telescoping building maintenance unit

Preliminary analysis of the building using a simplified analytical model indicated that the tower would be wind-sensitive and accelerations could be excessive. The simplified model comprised the central core as a cantilever linked to the outer frames, with axially stiff linkages representing the floors, the entire assemblage being considered as a plane frame. Having gained considerable insight into the behavior of the structure from the preliminary analysis, the tube-in-tube structural system was chosen for resistance to lateral wind loads.

During the preliminary design stage a 1:400 aeroelastic model was being developed and tested in a wind tunnel to determine and minimize wind pressures by varying the dynamic parameters. Considerable analytical work was carried out to tune the structure aeroelastically. The stiffness and mass of various structural components were adjusted and readjusted in this process to minimize the aeroelastic forces.

Once the structural form was finalized, a rigorous three-dimensional tube-in-tube analysis was carried out. This was necessary to ensure that displacements and accelerations under wind loading were below acceptable levels. In the analysis for core-frame interaction, the structure was propped at the ground floor and at each of the basement levels so that lateral loads could be transferred out to the site perimeter walls through diaphragm action of the floor slabs. Propping of the structure at the ground floor and basements avoided the problem of having to deal with large moments at the core footing and also served to control deflections and accelerations of the building under wind load. Of particular importance was the cross-wind response of the building, which produced a resulting moment 1.6 times the along-wind response.

The central core occupies a space approximately 16 m (52.5 ft) square in the center of the building and is, in reality, two cores with an elevator foyer space between. The two cores are linked together via floor slabs and beams, and in addition, by large diaphragms in the atrium and plant rooms. The atrium diaphragms were found to be particularly effective in reducing deflections by giving the building an exceptionally high point of rotation approximately 45 m (148 ft) above street level.

The central core is a multicell reinforced concrete structure with wall thicknesses varying from 200 to 600 mm (8 to 24 in.). Reinforcement ratios vary from about 1% in the lower parts of the building to 0.5% at the top. The core was designed globally for biaxial bending and axial load using the program FAILSAFE. In this program a particular section of the core is defined as an assembly of square elements within a system of coordinates, and the quantity and location of steel is also defined within the coordinate system. The program outputs a failure surface for axial load versus moment.

A detailed design of the core at headers, coupling beams, and diaphragms was carried out using deep-beam theory, shear-friction theory, and conventional reinforced concrete theory, as appropriate for the element under consideration.

Basement floors were designed as conventional reinforced concrete flat slabs, except that two special effects required particular attention in the design and detailing of reinforcements, namely, (1) transfer of wind loads out of the core to the basement walls, and (2) differential settlement between the core, major columns, and basement columns. Particular attention was paid to detailing the reinforcements at the core-slab joints, both on the drawing board and on site during construction.

The ground-floor slab was designed in reinforced concrete, incorporating an extensive beam system. At this level the wind-propping loads were considerably higher than in the basement slabs, and in addition the slab was designed to support a 10-kPa (200-psf) construction live load to allow for scaffolding up to support level 4 plant-room slab over the atrium.

The ground-floor slab is a multilevel slab with sloping and stepped portions, and in the northeast corner it contained large openings. Special bands of heavy reinforcing steel were required around the perimeter to transfer wind loads into perimeter walls. A diagonal band of heavy steel from the core to the northwest corner of the site was required to ensure a load path to compensate for the large penetrations of the northeast corner.

Tower floors were designed as posttensioned flat plates spanning approximately 10 m (33 ft) from the spandrel beams to the central core. Typical floor slabs are 275 mm (11 in.) thick and are stressed with tendons in bands of six, each tendon comprising five 12.7-mm (0.5-in.)-diameter supergrade strands in 90-mm (3.5-in.)-wide ducts. The banded tendon arrangement provides maximum flexibility of floor layout for the positioning of penetrations for services and internal stairs in the tenancy design stage.

The flat-plate soffit was important in allowing the builder to speed up the formwork placing and in achieving the specified cycle times. Posttensioning also meant minimum passive reinforcement, another feature to assist the builder.

Finite-element analysis of the floor slab indicated the existence of high shear stresses near the corners of the core. This was dealt with by installing some shear steel locally in the slab near each corner of the core. Spandrel beams were generally reinforced concrete, except for the longer cantilever beams at the corners of the building, which were posttensioned to minimize deflections.

3.3 CONDENSED REFERENCES/BIBLIOGRAPHY

Kilmister 1983, *Design and Construction of the Luth Headquarters Building, Kuala Lumpur.*
Martin 1989, *Wind Design of Four Buildings up to 306 m Tall.*
L'indústria Italiàna del Cemènto 1987, *The Luth Building in Kuala Lumpur (Malaysia)*

4

Lateral Load
Resisting Systems

4.1 BRACED FRAME AND MOMENT RESISTING FRAME SYSTEMS

Two fundamental lateral force resisting systems are the braced frame (also known as shear truss or vertical truss) and the moment resisting frame (moment frame or rigid frame). These systems evolved during the beginning of high-rise construction in the early twentieth century. Braced frames and moment resisting frames are normally organized as planar assemblies in orthogonal directions to create planar frames or a tube frame system. The two systems may be used together as an overall interactive system, thereby extending their individual applications to taller buildings. Both systems are commonly used today as effective means of resisting lateral forces in high-rise construction for buildings of up to 40 or 50 stories.

1 Braced Frames

Braced frames are cantilevered vertical trusses resisting lateral loads primarily through the axial stiffness of the frame members. Axial shortening and elongation of the column members under lateral loading accounts for 80 to 90% of the overall system deformation for slender truss systems. The effectiveness of the system, as characterized by a high ratio of stiffness to material quantity, is recognized for multistory buildings in the low- to midheight range.

Braced frame geometries are grouped, based on their ductility characteristics, as either concentric braced frames (CBF) or eccentric braced frames (EBF). In CBFs the axes of all members intersect at a point such that the member forces are axial. CBFs have a great amount of stiffness but low ductility. Thus in areas of low seismic activity, where high ductility is not essential, CBFs are the first choice of engineers for lateral load resistance. EBFs, on the other hand, utilize axis offsets to introduce flexure and shear into the frame, which lowers the stiffness-to-weight ratio but increases ductility.

The CBF can take the form of an X, Pratt, diagonal, K, or V, as shown in Fig. 4.1. The X bracings exhibit higher lateral stiffness-to-weight ratios in comparison to K or V bracings. However, the X bracings create a short circuit in the column gravity load transfer path as they absorb a portion of the column load in proportion to their stiffness. This creates additional forces in both diagonal and horizontal members of X-bracing systems which need to be considered in system design.

To accommodate door and other openings, EBFs are commonly used, as shown in Fig. 4.2. The shear and flexural action caused by the axis offset in the link beam improves ductility. Higher ductility through inelastic shear or bending action of the link beam make it a desirable lateral system in areas of high seismic activity. Ductility is measured by a well-behaved hysteresis loop and achieved through proper connection and member design such that all modes of instabilities and brittle failures are eliminated.

Braced frames are most often made from structural steel because of ease of construction. Depending on the diagonal force, length, required stiffness, and clearances, the diagonal member in structural steel can be made of double angles, channels, tees, tubes, or wide-flange shapes. Besides performance, the shape of the diagonal is often based on connection considerations. Examples of typical braced frame connections are depicted in Fig. 4.3.

Vertical trusses are often located in the elevator and service core areas of high-rise buildings, where frame diagonals may be enclosed within permanent walls. Braced frames can be joined to form closed section cells, which together are effective in resisting torsional forces. These cells may be bundled to take advantage of additional stiffness and provide a systematic means of dropping off the cells at the upper levels of a

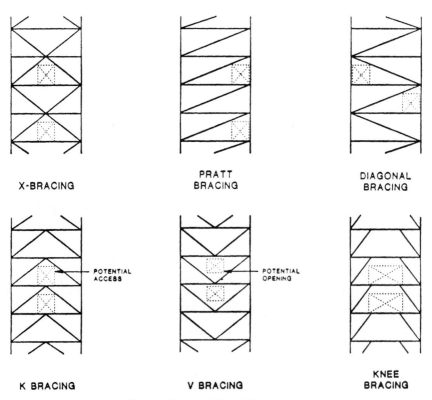

Fig. 4.1   Concentric braced frame forms.

tower where lateral forces are reduced. The strength and stiffness of the truss system is thus sensitive to the footprint of the core area and the arrangement of the elevators.

When the slenderness ratio of a core truss (the ratio of truss height to least width) increases, the overall overturning effect manifests itself in increased axial deformation and uplift forces of chord columns. While truss chord members may readily be designed to resist tension forces, net foundation uplift forces are generally undesirable. A design objective would be to spread the chords as far apart as possible while diverting gravity load to these chords to prevent or reduce the net tensile force.

As slenderness increases, the axial deformations of the chord columns of a truss system become more critical in controlling the sway of the structure. Increasing the stiffness and strength of the chord members in proportion to the work done by those members will provide an effective way to minimize sway. The bracing system between the chords can be designed to transfer the gravity loads of any intermediate chord columns to the boundary chord columns. As a result the intermediate chord columns could be eliminated or minimized in size and the efficiency of the boundary chords maximized.

To further reduce the steel tonnage and cost of the structure, composite steel and concrete chord columns may be utilized. Using concrete in chord columns will most likely provide a lower unit price for strength and axial stiffness.

2 Moment Resisting Frames

The moment resisting frame consists of horizontal and vertical members rigidly connected together in a planar grid form which resists lateral loads primarily through the flexural stiffness of the members. Typical deformations of the moment resisting frame system under lateral load are indicated in Fig. 4.4. A point of contraflexure is normally located near the midheight of the columns and midspan of the beams. The lateral deformation of the frame is due partly to the frame racking, which might be called shear sway, and partly to column shortening. The shear-sway component constitutes approximately 80 to 90% of the overall lateral deformation of the frame. The remaining portion of deformation is due to column shortening (cantilever component or so-called chord drift).

Moment resisting frames have advantages in high-rise construction due to their flexibility in architectural planning. A moment resisting frame may be placed in or around the core, on the exterior, or throughout the interior of the building with minimal constraint on the planning module. The frame may be architecturally exposed to express the gridlike nature of the structure. The spacing of the columns in a moment resisting frame can match that required for gravity framing. In fact the steel weight premium for lateral frame resistance decreases with increasing gravity loads on the frame.

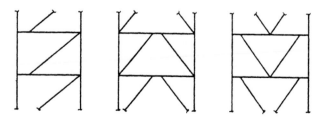

Fig. 4.2   Eccentric braced frame forms.

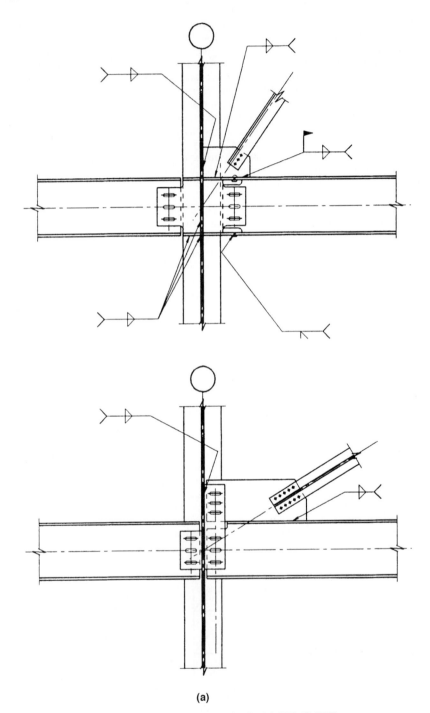

(a)

Fig. 4.3   Typical connection details. (*a*) **CBF.** (*b*) **EBF.**

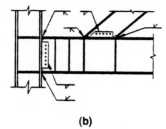

(b)

Fig. 4.3   Typical connection details. (*a*) CBF. (*b*) EBF. (*Continued*)

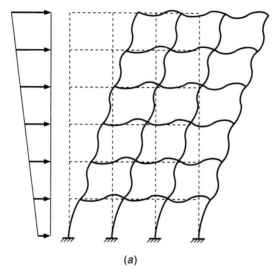

(*a*)

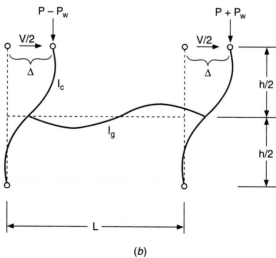

(*b*)

Fig. 4.4   **Moment resisting frame deformation under later load. (*a*) Frame deformation. (*b*) Frame behavior.**

The size of the members in a moment resisting frame is often controlled by stiffness rather than strength in order to develop acceptable drift control under lateral loads. The lateral drift is a function of both the column stiffness and beam stiffness. As beam spans are often greater than floor heights, the beam inertia needs to be greater than the column inertia by the ratio of beam span to story height for an effective moment resisting frame. Efficient floor framing would divert gravity load to the moment resisting frame until the gravity framing sizes match the size of the members required for stiffness. Moment resisting frame members, effectively sized for lateral sway control, increase in size and stiffness from top to bottom in proportion to the lateral shear. For gravity loading, the column sizes increase from top to bottom in the same fashion as a moment resisting frame, but the beam sizes remain constant. Based on this consideration, it would not be possible to match beam sizes for gravity and lateral load requirements. At best, one could try to minimize the difference.

Moment resisting frames are normally efficient for buildings no more than 20 or 30 stories in height. The lack of efficiency for taller buildings is due to the moment resisting frame's reliance on lateral load resistance derived primarily through flexure of its members. Member proportions and material costs become unreasonable for buildings higher than 20 or 30 stories, as is evident in the chart of steel quantity versus building height in Fig. 4.5.

Advances in the development of computer systems have allowed for the ready analysis of the highly indeterminate moment resisting frame. Effective guidelines and analysis procedures have been developed for the preliminary selection of moment resisting frame members. Optimization techniques are available which will determine the most efficient distribution of material in a moment resisting frame for a given deformation limit analysis.

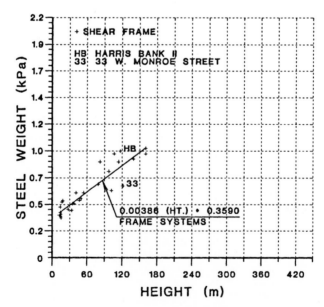

Fig. 4.5   Steel quantity versus height for frame systems.

The connections in the structural steel moment resisting frame are important design elements. Joint rotation can account for a significant portion of the lateral sway. The strength and ductility of the connection are also important considerations in both design and construction, especially for frames designed to resist seismic loads. It is important to understand the resources and practices of the local marketplace in order to design a connection that meets performance assumptions and is economical to construct.

Steel moment resisting frames may be formed by either bolted or fully welded joints. Construction of a fully rigid welded connection with the required strength, ductility, and reliability is readily achieved. Fully rigid bolted connections are commonly used where field welding is not practical. Semi-rigid connections are being used where joint moment-rotation properties have been included in the analysis. Compared to rigid connections, semi-rigid connections are easier to construct and more cost-effective, but their performance characteristics limit their use to low-rise buildings. Examples of rigid and semi-rigid connections are shown in Fig. 4.6.

Reinforced concrete frames have the advantage of a monolithically constructed joint ideally suited to the moment resisting frame system. Recent developments in concrete moment resisting frames include improved concrete characteristics, reinforcement detailing to provide improved ductility, and frame forming techniques.

The most recent advances in moment resisting frames have occurred in steel-concrete composite organizations. By mixing both steel and concrete, a wide variety of opportunities becomes available. The typical system composition normally involves light steel members, allowing rapid early erection of the frame, which are later encased in concrete and reinforcement, providing the necessary stiffness and strength for the finished structure. The advantages of both steel and concrete can be combined to achieve a superior system. Damping and axial strength can be improved by the addition of concrete to a steel frame. Reliability and ductility can be improved by the inclusion of steel shapes in a concrete frame. In a composite moment resisting frame, the beneficial properties of concrete are maintained without sacrificing the erection advantages of steel.

In Japan the moment resisting frame was first used in the 1960s. It was not until then that seismic technology had sufficiently advanced to allow the use of moment resisting frames. Frames subject to seismic loads must have sufficient joint strength and ductility as well as member stability when loaded beyond elastic conditions. Moment resisting frames must maintain their properties under cyclic loading beyond yield. In comparison to other lateral force resisting systems, moment resisting frames are well suited for use in seismic areas due to their ability to perform in the inelastic range. Further, moment resisting frames have a high degree of redundancy and can continue to perform even if one or more members fail. When moment resisting frames are used for seismic loadings, the columns should be designed to be stronger than the beams. This so-called strong-column–weak-beam theory ensures stability during plastic seismic overload. Current codes require special detailing to account for this requirement and designate such detailed frames as special moment resisting frames (SMRF).

3 Frame-Truss Interacting Systems

Vertical trusses alone may provide resistance for buildings of up to about 20 stories depending on the height-to-width ratio of the system. Shear trusses, when combined with moment resisting frames, produce a frame-truss interacting system. The linear wind sway of the moment frame, when combined with the cantilever parabolic sway of the truss, produces enhanced lateral stiffness. The truss is restrained by the frame at the upper part of the building, whereas at the lower part, the truss restrains the frame. This involves a transfer of shear forces from the top to the bottom of the building. Figure 4.7 shows the truss and frame deflections if each resisted the full wind shear. The distribution of wind shear between truss and frame can also be noted. Frame-truss interacting systems have a wide range of application to buildings of up to 40 stories in height.

In general, core trusses are combined with moment frames located on the building perimeter, where the column spacing and the member proportions of the frame may be appropriately manipulated. Optimum efficiency is obtained when gravity-designed columns are used as truss chords without increasing them for wind forces. These are then combined with gravity-designed exterior columns and spandrel beams with rigid

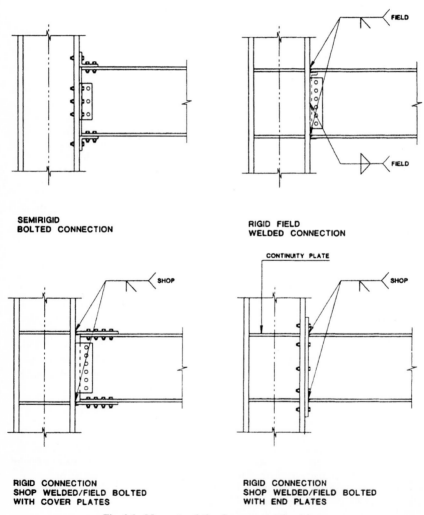

SEMIRIGID
BOLTED CONNECTION

RIGID FIELD
WELDED CONNECTION

RIGID CONNECTION
SHOP WELDED/FIELD BOLTED
WITH COVER PLATES

RIGID CONNECTION
SHOP WELDED/FIELD BOLTED
WITH END PLATES

Fig. 4.6   Moment resisting frame connection types.

connections. If the lateral stiffness of the system is adequate, this then would produce an optimal design. If additional stiffness is required, the decision of whether to increase the core or the frame members depends on the relative efficiency of the two components. The frame beam spans, story heights, and core truss depth are key parameters. Tension or uplift conditions may limit the possibility of increasing chord columns.

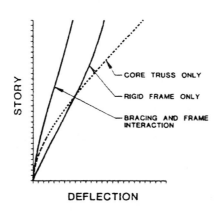

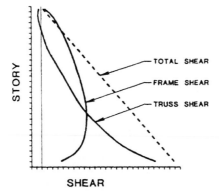

Fig. 4.7   Frame-truss interacting system.

PROJECT DESCRIPTIONS

Sanwa Bank
Tokyo, Japan

Architect	Nikken Sekkei Ltd.
Structural engineer	Nikken Sekkei Ltd.
Year of completion	1973
Height from street to roof	99.7 m (327 ft)
Number of stories	25
Number of levels below ground	4
Building use	Office
Frame material	Structural steel
Typical floor live load	3 kPa (60 psf)
Basic wind velocity	Not available
Maximum lateral deflection	Not available
Design fundamental period	3 sec in both directions
Design acceleration	20 mg; 40 mg for seismic loading
Design damping	2% of critical
Earthquake loading	$C = 0.10$
Type of structure	Combination of rigid frames and eccentric K bracing
Foundation conditions	Alluvium and diluvial gravel
Footing type	Raft on reinforced concrete driven piles
Typical floor	
Story height	3.84 m (12 ft 6 in.)
Beam span	24 m (78 ft 9 in.)
Beam depth	850 mm (33.5 in.)
Beam spacing	3.15 m (10 ft 4 in.)
Material	Steel, grade 400 MPa (58 ksi) 2d floor and above; concrete-encased steel below 2d floor
Slab	120-mm (4.75-in.) reinforced concrete
Columns	
Size at ground floor	400- by 400-mm (16- by 16-in.) H sections
Spacing	3.15 m (10 ft 4 in.)
Material	Steel, grade SM 490, 483 MPa (70 ksi)
Core	Shear walls below 2d floor, 800 mm (31.5 in.) thick; combined rigid and braced steel frames, grade SM 490, above 2d floor

In designing the Sanwa Bank building for earthquake and wind loads (Fig. 4.8), it was decided to place eccentric K-braced frames at appropriate locations such that they will act together with the rigid frames. The diagonal members of eccentric K-braced frames do not intersect at the center of the beam. Thus yielding at the center of the beams will occur before braces buckle, ensuring ductility and allowing for adjustment of the frame ductility (Figs. 4.9 to 4.11). This has been confirmed, both experimentally and theoretically.

Ductility and strength are ensured by using a composite beam for the 24-m (78.9-in.) office floor spans. This also minimizes vibration disturbance due to people walking, as was confirmed through a composite beam mock-up test.

Precast concrete panels faced with granite are used as cladding material, providing a solid appearance to the building (Fig. 4.12). The panel fixings were designed so that during an earthquake, the panels can follow the building deformations without damage or risk of dislodgement. This was checked using a two-story two-span full-scale model.

Fig. 4.8 Sanwa Bank, Tokyo, Japan.

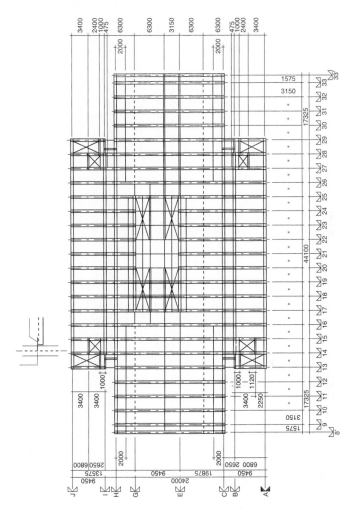

Fig. 4.9 Typical structural floor plan; Sanwa Bank.

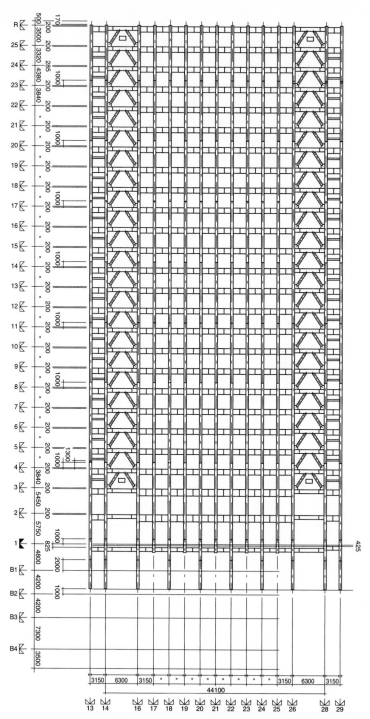

Fig. 4.10   Framework; Sanwa Bank.

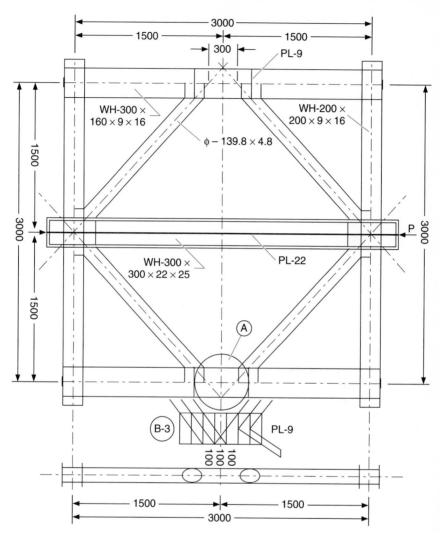

Fig. 4.11   Specimen of eccentric K frame; Sanwa Bank.

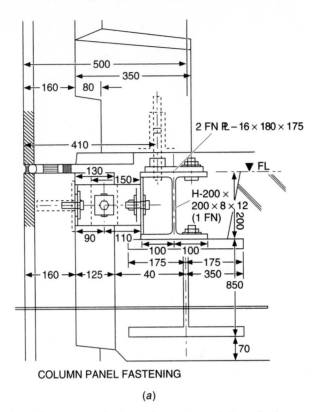

COLUMN PANEL FASTENING

(a)

Fig. 4.12   Details of precast concrete panel; Sanwa Bank.

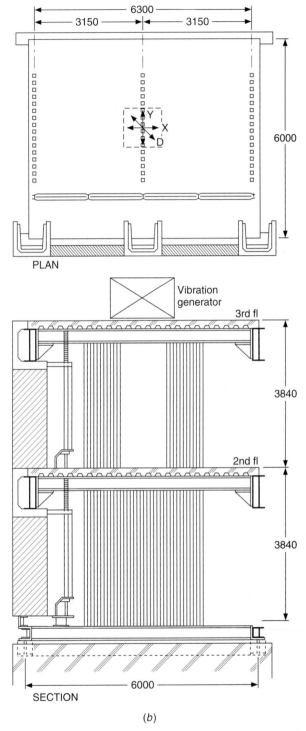

6300

3150 3150

6000

Y
X
D

PLAN

Vibration
generator

3rd fl

3840

2nd fl

3840

6000

SECTION

(b)

Fig. 4.12   Details of precast concrete panel; Sanwa Bank. (*Continued*)

ACT Tower
Hamamatsu City, Japan

Architect	Nihon Sekkei Inc. and Mitsubishi Estate Co. Ltd.
Structural engineer	Nihon Sekkei Inc.
Year of completion	1994
Height from street to roof	211.9 m (695 ft)
Number of stories	47
Number of levels below ground	2
Building use	Hotel, offices, retail space
Frame material	Steel
Typical floor live load	5 kPa (100 psf)
Basic wind velocity	30 m/sec (67 mph)
Maximum lateral deflection	$H/200$, 100-yr return period wind
Design fundamental period	4.52, 4.73 sec
Design acceleration	52 mg peak, 100-yr return period
Design damping	1% serviceability, 2% ultimate
Earthquake loading	$C = 0.06$
Type of structure	Braced frames
Foundation conditions	Clay, sand, and gravel
Footing type	Piles 1.5 to 2.4 m (5 to 8 ft) in diameter, 25 to 30 m (82 to 98 ft) long
Typical floor	
Story height	4 m (13 ft) office; 3.15 m (10 ft 4 in.) hotel
Beam span	17.5 m (57 ft 5 in.) max. office; 10 m (33 ft 10 in.) hotel
Beam depth	850 mm (33.5 in.) office; 700 mm (27.5 in.) hotel
Beam spacing	3.2, 6.4 m (10 ft 6 in., 21 ft) office; 3.2, 4.27 m (10 ft 6 in., 14 ft) hotel
Slab	135- to 180-mm (5.25- to 7-in.) concrete
Columns	
Size at ground floor	750 by 600 mm (30 by 24 in.)
Spacing	3.2 and 6.4 m (10 ft 6 in. and 21 ft)
Core	X- and K-braced frames

Braced frames were used to increase the stiffness of the ACT Tower (Fig. 4.13) and to achieve an optimum structural system (Figs. 4.14 to 4.16). Three wind-tunnel tests were performed:

1. A wind pressure test to evaluate facade pressures
2. A wind force test to measure the horizontal force, overturning moment, and torsional moment

3. A dynamic test to check the dynamic analysis results

The dynamic analysis was performed using the mean and the standard deviation as well as the power spectrum of the overturning moment and the torsional moment coefficients obtained in the wind force test.

The building response spectra are obtained by combining the wind spectra (for the x, y, and θ directions) and the magnification factors versus frequency curve. As the building cross section is ellipsoidal, special consideration was given to getting the maximum response values used in the design in the x, y, and θ directions. The dynamic stability and the possibility of galloping were also checked.

Strong winds can occur several times a year, causing uncomfortable building motion. In order to avoid this problem, a damping system has been installed to reduce the acceleration in the y direction.

The building site is located in a very active seismic area. The largest earthquakes in this zone to date were of magnitude 8. A special seismic analysis was performed using the data of the three largest earthquakes that have originated in this area in order to model the earthquake waves and the maximum possible accelerations for the ACT Tower site. These 3 earthquake waves were 416 gal/sec (550 mm/sec) (Ansei Tohka earthquake); 150 gal/sec (320 mm/sec) (Nohbi earthquake); and 332 gal/sec (850 mm/sec) (Tohnankai earthquake).

Fig. 4.13   ACT Tower, Hamamatsu City, Japan.

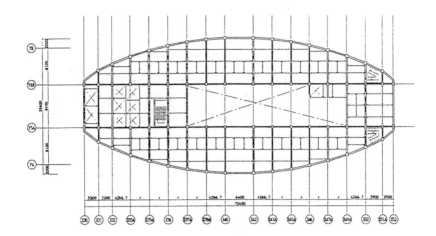

Typical Structural Plan (Hotel)

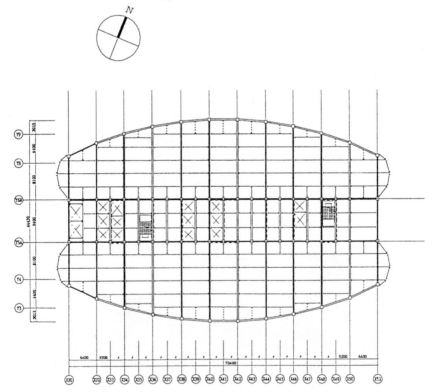

Typical Structural Plan (Office)

Fig. 4.14   Typical structural plans; ACT Tower.

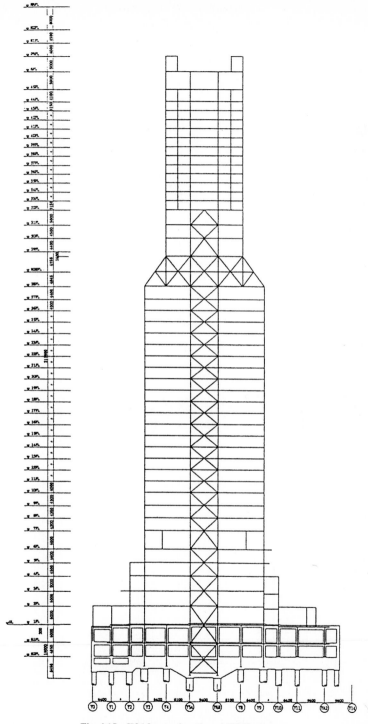

Fig. 4.15   X36 frame elevation; ACT Tower.

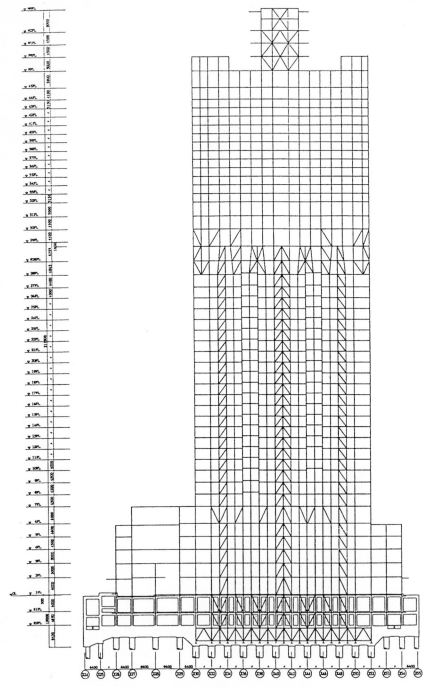

Fig. 4.16   Y5A frame elevation; ACT Tower.

Kobe Portopia Hotel
Kobe, Japan

Architect	Nikken Sekkei Ltd. with Portopia Hotel Design Office
Structural engineer	Nikken Sekkei Ltd. with Portopia Hotel Design Office
Year of completion	1981
Height from street to roof	112 m (367 ft)
Number of stories	31
Number of levels below ground	2
Building use	Hotel
Frame material	Structural steel
Typical floor live load	1.8 kPa (36 psf)
Basic wind velocity	Not available
Maximum lateral deflection	350 mm (13.75 in.)
Design fundamental period	3.5 sec transverse; 3.6 sec longitudinal
Design acceleration	20 mg; 35 mg for seismic loading
Design damping	2%
Earthquake loading	$C = 0.08$
Type of structure	Moment frame and braced frame
Foundation conditions	Fill over alluvial and diluvial strata
Footing type	Raft on prestressed concrete driven piles
Typical floor	
Story height	3.02 m (9 ft 11 in.)
Beam span	7.5 and 6.75 m (24 ft 7 in. and 22 ft 2 in.)
Beam depth	800 mm (31.5 in.)
Beam spacing	7.5 m (24 ft 7 in.)
Material	Steel, grade 400 and 490 MPa (58 and 70 ksi) 5th floor and above; concrete-encased steel below 5th floor
Slab	130-mm (5-in.) reinforced concrete
Columns	
Size at ground floor	1100 by 1100 mm (43 by 43 in.)
Spacing	7.5 m (24 ft 7 in.)
Material	Steel encased in 24-MPa (3400-psi) concrete
Core	600-mm (24-in.) concrete shear walls below 5th floor; structural steel rigid frames 5th floor and above

The typical floor plan of the Kobe Portopia Hotel (Fig. 4.17) is an oval, measuring 75.5 m (248 ft) in the east-west direction and 13.5 m (44 ft) in the north-south direction (Fig. 4.18). Above the fifth floor of the high-rise part, strength and ductility are provided by using a reinforced concrete rigid frame. The fifth and lower floors, which have a larger story height, have a composite structure of shear walls and rigid frames made of steel encased in reinforced concrete (Fig. 4.19).

The site is part of about 500 ha (1200 acres) of artificially reclaimed ground, which has been filled over a period of 10 years, starting in 1965. Before building construction commenced, the site was preloaded, theoretically completing settlement of the former 12-m (40-ft)-thick sea-bottom clay layer. Because the building weight is about 100,000

Fig. 4.17   Kobe Portopia Hotel, Kobe, Japan.

tonnes (110,000 tons), a basement with good foundation load balance was possible, with the weight of the excavated soil being designed to exceed the weight of the building.

Piles of about 40-m (130-ft) length were used. The building is supported by using the diluvial layer as the bearing stratum. In pile design, pile groups were used wherever possible to cope with unmeasured negative friction. Structural safety was confirmed by performing a seismic response analysis of the building-pile-bearing stratum composite form against horizontal seismic loads.

The floor plan has an unusual form, so various wind tunnel tests were performed to investigate such factors as the wind force coefficient, the wind pressure coefficient, ambient wind velocity, and the dynamic stability against wind. In everything from the structure itself to cladding materials, external doors and windows, and ground-level wind velocity, wind tunnel test results were used to ensure adequate safety and serviceability.

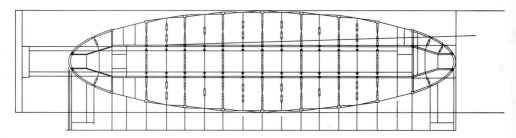

Fig. 4.18   Typical structural floor plan; Kobe Portopia Hotel.

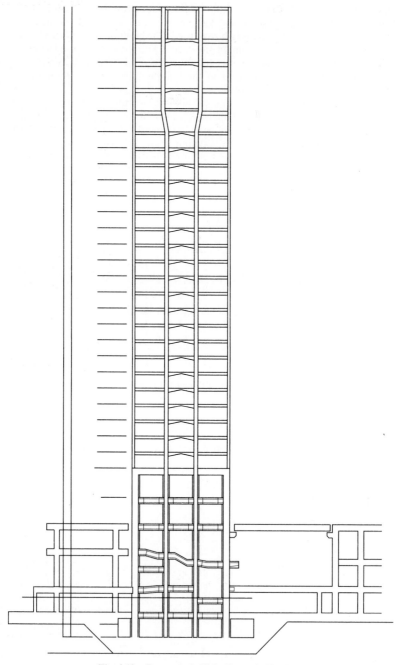

Fig. 4.19   Framework; Kobe Portopia Hotel.

Nankai South Tower Hotel
Osaka, Japan

Architect	Nikken Sekkei Ltd.
Structural engineer	Nikken Sekkei Ltd.
Year of completion	1990
Height from street to roof	147 m (482 ft)
Number of stories	36
Number of levels below ground	3
Building use	Hotel
Frame material	Structural steel upper floors; concrete-encased structural steel plus concrete shear walls lower floors
Typical floor live load	1.8 kPa (36 psf)
Basic wind velocity	35 m/sec (78 mph)
Maximum lateral deflection	Not available
Design fundamental period	3.24 sec transverse; 3.03 sec longitudinal
Design acceleration	Level 1 EQ, 13 to 25 mg; level 2 EQ, 21 to 40 mg
Design damping	2%
Earthquake loading	$C = 0.120$
Type of structure	Level 5 and above, rigid frames; level 4 and below, combined frames and shear walls
Foundation conditions	Gravel
Footing type	Cast-in-place 2-m (6.5-ft)-diameter bored piles 10 m (33 ft) deep
Typical floor	
Story height	3.2 m (10 ft 6 in.)
Beam span	Primary, 10.5 m (34 ft 5 in.); secondary, 5.4 m (17 ft 8 in.)
Beam depth	850 mm (33.5 in.)
Beam spacing	2.625 m (8 ft 7 in.)
Material	Steel, grade 400 and 490 MPa (58 and 70 ksi)
Slab	140-mm (5.5-in.) concrete on metal deck
Columns	
Size at ground level	1300 by 1300 mm (51 by 51 in.)
Spacing	10.5 m (34 ft 5 in.)
Material	Steel, grade 49 MPa (7000 psi)
Core	Shear wall, 34-MPa (3400-psi) concrete, 350 mm (14 in.) thick

This hotel was constructed over a railway station, which had been designed and constructed by another firm up to the fourth floor 10 years earlier (Fig. 4.20). An expansion
of about the same extent was planned even in the original design, but there existed limitations with regard to the allowable stress of the already constructed parts, including the
piles. While over the course of 10 years structural codes had been modified, making it
more difficult to expand buildings constructed before the code changes, the design tech-

Fig. 4.20   Nankai South Tower Hotel, Osaka, Japan.

niques for high-rise buildings had not fundamentally changed, so the strength of the already constructed parts was for the most part adequate. However, there was a planning regulation change in that guest rooms must now have balconies, and it was necessary to comply with the desires of a designer, which changed the plans considerably.

The increased weight due to balconies was handled by changing the specific gravity of the concrete from an original 1.8 to 1.65. In the original design, slanted columns had ranged from the sixth to the twentieth floors, which was due to changes in the spans of the upper and lower floors, and the designer wanted to reduce this range to between floors 9 and 12. To improve the ensuing reduction in building rigidity, the size of the external columns was increased. This suppressed the overall bending deformation, and at the same time the inner columns were effectively used as shear columns. External columns are large boxed members, so in the length direction the perimeter frame is used to resist all of the horizontal loading (Figs. 4.21 and 4.22).

To facilitate construction, balconies were designed in the L shape with a length of 10.5 m (34 ft 5 in.). Prestressed concrete, only 90 mm (3.5 in.) thick, was used to minimize the weight.

A composite floor, fire rated for 2 hours, was used in the typical guest room. The deck has to be of the linked beam type (which covers at least two beam spans). In unit bath areas, which had to be partially dropped, ordinary slabs using a flat deck were employed.

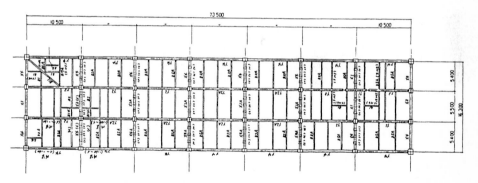

Fig. 4.21   Typical floor plan; Nankai South Tower Hotel.

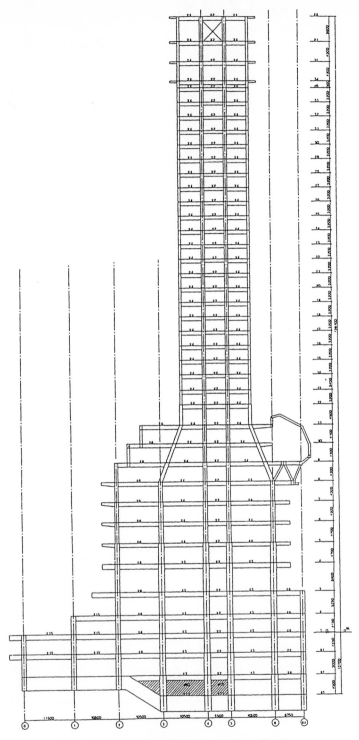

Fig. 4.22   Framework; Nankai South Tower Hotel.

World Trade Center
Osaka, Japan

Architect	Nikken Sekkei Ltd. with Mancini Duffy Associates
Structural engineer	Nikken Sekkei Ltd. with Mancini Duffy Associates
Year of completion	1994
Height from street to roof	252 m (827 ft)
Number of stories	55
Number of levels below ground	3
Building use	Office
Frame material	Structural steel
Typical floor live load	3 kPa (60 psf)
Basic wind velocity	40 m/sec (90 mph)
Maximum lateral deflection	1300 mm (51 in.), 200-yr return period wind
Design fundamental period	5.3 sec transverse; 5.8 sec longitudinal
Design velocity	Level 1 EQ, 250 mm/sec (10 in./sec); level 2 EQ, 500 mm/sec (20 in./sec)
Design damping	2%
Earthquake loading	$C = 0.05$ longitudinal; $C = 0.075$ transverse
Type of structure	Rigid frames with core braced transversely
Foundation conditions	20-m (65-ft 7-in.) fill over alluvial clay and sand strata
Footing type	Cast-in-place steel-lined bored piles belled at their base
Typical floor	
Story height	4.0 m (13 ft 1.5 in.)
Beam span	16 m (52 ft 6 in.)
Beam depth	900 mm (35 ft 5 in.)
Beam spacing	3.2 and 9.6 m (10 ft 6 in. and 31 ft 6 in.)
Material	Steel, grade 400 and 490 MPa (58 and 70 ksi)
Slab	175-mm (7-in.) concrete on metal deck
Columns	
Size at ground floor	650 by 850 mm (25.5 by 33.5 in.)
Spacing	3.2 and 9.6 m (10 ft 6 in. and 31 ft 6 in.)
Material	Steel, grade SM 53B
Core	Steel frames, braced in transverse direction
Material	Steel, grade 490 MPa (70 ksi)

This 252-m (827-ft)-high building stands on reclaimed land in the Osaka Nanko (south port) area (Fig. 4.23). As a consequence, the design of the foundation structure and the resistance to wind were painstakingly investigated.

A typical high-rise floor is 36 by 70 m (118 by 230 ft), and the building has an extremely slender form where the ratio of short side to height is 1:7 (Figs. 4.24 and 4.25). Below the seventh floor, columns are transferred to the perimeter, forming a supertruss frame in order to strengthen the resistance to wind and earthquakes, and widely distribute axial forces of the high-rise building over the ground. This forms a "base" for the tower, which is integrated with the underground structure.

Wind is a more dominant lateral load for this building than earthquakes. The wind load for the design, including vibration assessment, was determined from the results of wind tunnel testing. The testing investigated instabilities as well as accelerations likely to affect the comfort of occupants, unstable vibration due to wind, and habitability during swaying of the building due to wind forces.

As the site is artificially reclaimed land, and settlement due to earth filling is not complete, the cast-in-place steel-pipe concrete piles used are coated with asphalt to reduce friction with the surrounding ground. The bearing stratum is a diluvial sandy gravel layer around 60 m (197 ft) below ground level.

Fig. 4.23 World Trade Center, Osaka, Japan.

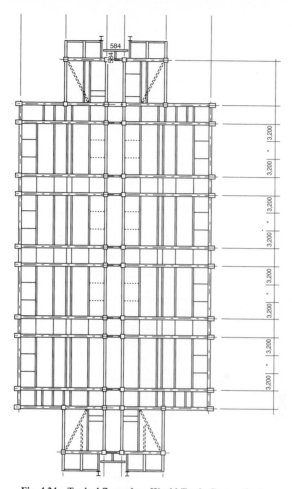

Fig. 4.24   Typical floor plan; World Trade Center, Osaka.

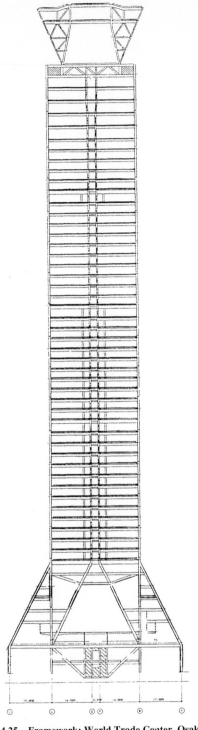

Fig. 4.25   Framework; World Trade Center, Osaka.

84

Kobe Commerce, Industry and Trade Center
Kobe, Japan

Architect	Nikken Sekkei Ltd.
Structural engineer	Nikken Sekkei Ltd.
Year of completion	1969
Height from street to roof	110.06 m (363 ft)
Number of stories	26
Number of levels below ground	2
Building use	Office
Frame material	Steel
Typical floor live load	3 kPa (60 psf)
Basic wind velocity	Unknown
Maximum lateral deflection	$H/400$
Design fundamental period	3.42 sec each direction
Design acceleration	20 mg elastic; 40 mg elastoplastic
Design damping	2%
Earthquake loading	$C = 0.085$
Type of structure	Perimeter framed tube with diagonally braced core
Foundation conditions	Alternating gravel and diluvial clay strata
Footing type	Raft
Typical floor	
Story height	3.84 m (12 ft 7 in.)
Beam span	9.45 m (31 ft)
Beam depth	600 mm (24 in.)
Beam spacing	3 m (9 ft 10 in.)
Material	Steel, grade 400 MPa (58 ksi) above 1st floor; concrete-encased structural steel 1st floor and below
Slab	160-mm (6.25-in.) concrete on metal deck
Columns	
Size at ground floor	700 by 700 mm (27.5 by 27.5 in.)
Spacing	3 m (9 ft 10 in.)
Material	Steel, grade 490 MPa (70 ksi)
Core	Structural steel with prestressing-bar diagonal bracing

This building is structurally characterized by its "tube-in-tube structure," which consists of perimeter wall frames with 3-m (10-ft) spans and internal braced frames using prestressing steel bars for diagonal bracing (Figs. 4.26 to 4.28). For the purpose of efficiently increasing the earthquake resisting capacity of a building, it is preferable to design its structure in a bending failure mode so as to disperse the yielding of frames during an earthquake. To achieve this objective, the tube-in-tube structure was adopted for this building.

For the braced frames using prestressing steel bars, F13T steel bars serve as diagonal braces (Fig. 4.29). These braces have a wide elastic range and thus can resist the maximum seismic forces within the elastic region. This enables the overall structure to act in a bending failure mode, thereby securing stable recovery characteristics. In this way the structure is designed to be effective from an aseismic viewpoint.

Fig. 4.26   Kobe Commerce, Industry and Trade Center, Kobe, Japan.

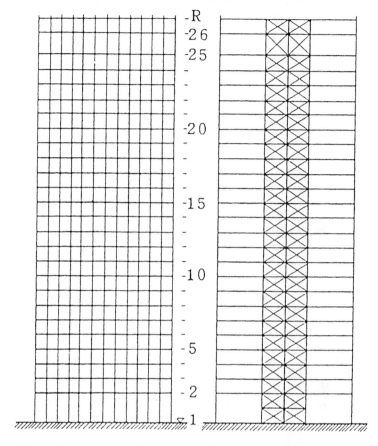

Perimeter frame   Braced frame

Fig. 4.27   Framework; Kobe Commerce, Industry and Trade Center.

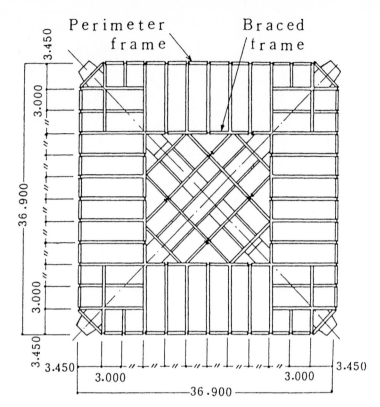

Fig. 4.28   Typical structural floor plan; Kobe Commerce, Industry and Trade Center.

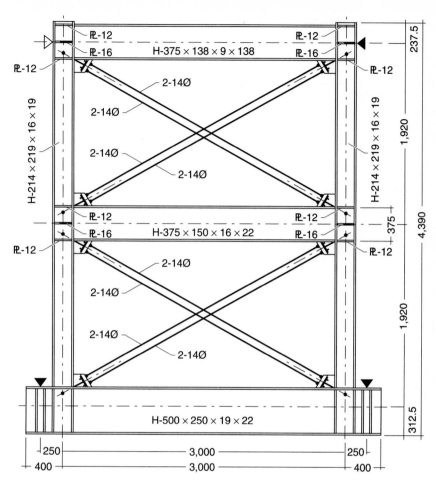

Fig. 4.29   **Specimen of braced frame using prestressing bars; Kobe Commerce, Industry and Trade Center.**

Marriott Marquis Hotel
New York, N.Y., USA

Architect	John Portman Associates
Structural engineer	Weidlinger Associates
Year of completion	1985
Height from street to roof	175 m (574 ft)
Number of stories	50
Number of levels below ground	2
Building use	Hotel
Frame material	Steel
Typical floor live load	2 kPa (40 psf)
Basic wind velocity	36 m/sec (80 mph)
Maximum lateral deflection	300 mm (12 in.), 100-yr return
Design fundamental period	5 sec
Design acceleration	20 mg peak, 10-yr return
Design damping	1% serviceability; 2% ultimate
Earthquake loading	Not applicable
Type of structure	Braced and rigid frames
Foundation conditions	Rock, 4-MPa (40-ton/ft^2) capacity
Footing type	Spread footings
Typical floor	
Story height	3.05 m (10 ft)
Beam span	8.53 m (28 ft)
Beam depth	460 mm (18 in.)
Beam spacing	3.05 m (10 ft)
Material	Steel, grade 250 MPa (36 ksi)
Slab	Precast concrete, 300 mm (12 in.) thick
Columns	610- by 610-mm (24- by 24-in.) built-up I shape from 90- to 203-mm (3.5- to 8-in.) plates, grade 30- to 35-MPa (4200 to 5000 psi) steel
Core	Reinforced concrete beam and column frame with 13 columns in a circle

Facing Times Square on a block front between 45th and 46th Streets, the new 167,000-m^2 (1.8 million ft^2) hotel rises 50 stories above the street (Fig. 4.30). The two sheer fin walls along the two side streets contrast sharply with the stepped and skylit facade facing Broadway. It is surmounted by a projecting, rotating cocktail lounge seven stories above the ground, actually the lobby level of the hotel. Above are five-story packages of hotel rooms that are stepped back and forth between the fin walls like a giant's ladder. The first six floors of the building contain public facilities, including a 1500-seat theater, a ballroom, exhibition and meeting rooms, and retail space.

Fig. 4.30   **Marriott Marquis Hotel, New York, under construction.** (*Photo by Jennifer Levy.*)

A circular concrete core, with 12 Tivoli lighted elevators and four enclosed elevators, rises from the street level through the public levels, breaking free at the lobby level, into a spectacular 35-story atrium (Fig. 4.31). It terminates at a multilevel rotating rooftop restaurant. Skylights on the east facade, between the five-story packages of hotel rooms, bring daylight into the atrium, shining down onto the hotel lobby. The 35 guest room floors, with 1876 rooms, are disposed in rectangular bands around the atrium. From the guest floor corridors, with their projecting planters presenting an image of the hanging gardens of Babylon, one can look down at the parklike lobby surrounded by colorful restaurants.

As a structure, the building is equally unique, consisting of a steel-framed structure surrounding the slip-formed concrete core. Between the two 11-m (36-ft)-deep fins, a 34-m (112-ft) clear span is framed using girders below the lobby and five-story Vieren-

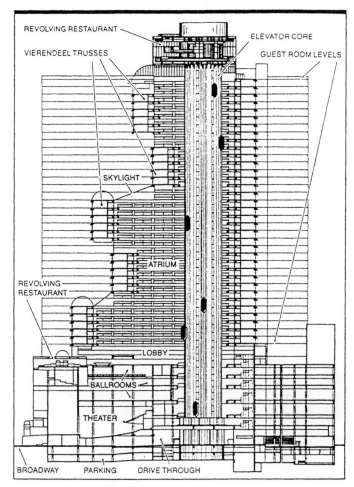

Fig. 4.31   Elevator core rises 180 m (600 ft) through atrium to revolving restaurants; Marriott Marquis Hotel.

deel frames for the packages of hotel rooms. These hotel room packages were originally conceived to be supported by steel trusses. The Vierendeel frames not only eliminated the trusses, but being tied into the side-wing vertical frames, provide stiffness in the north-south direction. Structurally, the building is a U with columns 8.5 m (28 ft) on center along the three sides, with the closure provided by the 34-m (112-ft)-span Vierendeel frames. The two sides of the building and the back are rigid frames above the lobby and are trussed between lobby and ground floor. To provide the required lateral stiffness in the front, the Vierendeels, combined with a subdivided vertical frame on the two sides, form superframes. In order to avoid the added columns at the ground level, the columns placed 6 m (20 ft) from the north or south side are, in fact, posts with vertical slip joints at midheight between floors.

At the back of the building, along column line 12, a single frame, cross braced below the lobby, provides the stiffness in the north-south direction. Since the group of three superframes in the front (at column lines 3, 4, and 5) have substantially different stiffness, a process of fine-tuning was undertaken to match deflections between the packet of superframes and the single frame as closely as possible. The purpose of this exercise was to avoid torsion in the building. In this connection it should be noted that even if the two were perfectly matched, a 5% eccentricity is required by the Uniform Building Code (UBC) between the centroids of mass and rigidity. This has the effect of requiring a 10% increase in shear carried by the diaphragm above that, resulting from lateral wind forces.

Below the lobby, the floor construction is conventional metal deck and concrete fill. For the guest room floors, this construction was originally specified. However, since a ceiling is required and since spans for the metal deck are limited, necessitating more beams, an alternative, using long-span precast concrete plank without topping, was chosen, based on economic considerations. Not only did this result in a reduction in the number of steel beams, but it also eliminated a hung ceiling since the underside of the plank is a finished surface. The more than 93,000 m^2 (1 million ft^2) of plank used makes this a most dramatic application of plank floors in a high-rise building. As a result of the innovative use of both the Vierendeel trusses and the concrete floor plank (which are only marginally heavier than the original metal deck and concrete solution), the steel structure with less than 117 kg/m^2 (24 psf) is extremely efficient and economical.

The planks, apart from providing normal vertical load-carrying capacity, are required to provide the diaphragm resistance, transferring all lateral forces to the vertical wind frames. Because of the height of the building and the unusual configuration, this implied special requirements for the plank. Basically, the plank must do the following:

1. Support dead and live loads
2. Transfer wind forces to bracing members
3. Transfer column-stability forces to bracing members
4. Transfer forces between bracing members

Since the planks are an inherent part of the stability of the structure, planks were placed, grouted, and welded in sequence with the erection of the steel frame. A rapid-setting, nonshrink grout with high early strength was specified for the grouting of the joints between planks. These joints, which have shear keys with castellations, have been shown by experiment to provide adequate shear strength for diaphragm action with a grout strength of 17 MPa (2500 psi). For this project, a design strength of 35 MPa (5000 psi) was specified to provide higher early strength and a margin of safety for the extreme weather conditions to be encountered during the construction cycle. Since no topping is used, diaphragm action relies solely on the integrity of the joint and the anchor.

Taj Mahal Hotel
Atlantic City, New Jersey, USA

Architect	Francis Xavier Dumont
Structural engineer	Paulus Sokolowski and Sartor, Inc.
Year of completion	1990
Height from street to roof	128 m (420 ft)
Number of stories	42
Number of levels below ground	0
Building use	Hotel
Frame material	Steel with precast concrete perimeter beams
Typical floor live load	Rooms 2 kPa (40 psf); corridors 4 kPa (80 psf)
Basic wind velocity	40 m/sec (90 mph)
Design wind load deflection	254 mm (10 in.)
Design fundamental period	3.0 sec
Design acceleration	1.8 mg rms, 1-yr return period
Design damping	1% serviceability
Earthquake loading	$C = 0.037$; $K = 1.0$ (did not govern design)
Type of structure	Staggered steel trusses and braced steel core transverse, rigid perimeter frame longitudinal direction
Foundation conditions	27 m (90 ft) of loose sand and thin organic strata over dense sand
Footing type	355-mm (14-in.)-diameter steel-shell driven cast-in-place concrete piles, 1500-kN (165-ton) capacity
Typical floor	
Story height	2.69 m (8 ft 10 in.)
Truss span	20.7 m (68 ft)
Truss depth	2.69 m (6 ft 10 in.)
Truss spacing	9.14 m (30 ft)
Material	Structural steel
Slab	100-mm (4-in.) precast slabs with 100-mm (4-in.) cast-in-place topping; 35-MPa (5000 psi) concrete
Columns	
Size at ground floor	Built-up steel, 2230 kg/m (1500 lb/ft)
Spacing	9.1 by 20.7 m (30 by 68 ft)
Material	Structural steel, grade 350 MPa (50 ksi)
Core	Braced steel frames, grade 350 MPa (50 ksi)

When the developer wanted a 1200-room high-rise luxury hotel right on the exposed oceanfront, a prime concern of the designers was occupant comfort. Building sway and acceleration had to be minimized.

Preliminary analyses and cost studies were made of four basic structural systems:

- Steel staggered truss with concrete floors
- Concrete frames with shear core and other shear walls
- Concrete-framed tube with concrete shear core
- Steel-framed tube with shear core

All systems except the staggered truss required relatively large shear walls to control sway, and the heavily loaded walls required large and expensive footing systems. In addition the steel-framed tube had problems of uplift. Relative costs were:

Steel staggered truss	1.00
Concrete frames and shear walls	1.25
Concrete-framed tube	1.10
Steel-framed tube	1.40

The staggered truss framing system was developed by a U.S. Steel–sponsored research team working at M.I.T. in the mid-1960s. Its basic element is the story-deep truss which spans the full width of the building at alternate floors on each column line. Hence the floor spans from the top chord of one truss to the bottom chord of adjacent trusses so that each truss is loaded on both the top and the bottom chords and is laterally fully restrained.

Because all gravity load is supported at the perimeter of the building, the whole of the building weight can be mobilized to resist overturning effects.

Lateral forces are transmitted from floor to floor down the building via the floor diaphragms and truss web members. The perimeter columns carry only axial load in the transverse direction and can therefore have their strong axis oriented longitudinally to form part of a longitudinal rigid frame.

The layout of the Taj Mahal Hotel (Fig. 4.32) with a central double-loaded corridor suited the staggered truss arrangement as it allowed the provision of a Vierendeel panel midspan, where the shear is least, for the corridor.

With the structural system selected, a wind tunnel study was carried out to determine structural forces, dynamic behavior, cladding pressures, and environmental effects at ground level. From this, the design lateral loading, building drift, and acceleration were established.

Because this is a tall building for a staggered truss system, the shears in the floors were an important design consideration. The 200-mm (8-in.)-thick slabs comprise precast pretensioned concrete planks tapered on their top surface from 127 mm (5 in.) thick midspan to 76 mm (3 in.) thick each end, where they are supported on top of the 254-mm (10-in.) -wide flange steel truss chord, and a cast-in-place topping. The shear connection between floors and trusses is achieved by stud shear connectors (Figs. 4.33 and 4.34).

Because of functional and architectural requirements, the staggered truss system could not be used at all locations. As a consequence, two other systems were used, a core frame and an end frame. The core frame consists of a truss system with a truss at every floor, but with a vertical stiffness to match the staggered trusses. The truss at every level was required to carry shears from lateral loading without reliance on the floor diaphragm which, at this location, is heavily penetrated by service openings.

The three-bay end frame comprises a diagonally braced center bay and outer bays of rigid framing connected to large perimeter columns. This frame is 13.7 m (45 ft) wide compared to the 20.7-m (68-ft) width of the typical frames (Fig. 4.35).

Wind shears are transferred to the foundations by embedding the bottom chords of the lowest trusses in large concrete-grade beams. On the lines where the lowest truss

Fig. 4.32   Taj Mahal Hotel, Atlantic City, New Jersey.

was one story above the footings, a diagonal brace was provided at the columns to transfer the load to a steel beam embedded in the footing, similar to the adjacent truss bottom chord. A pile cap at a typical staggered truss bay is 7.6 m (25 ft) square and 2.44 m (8 ft) deep, supported on 36 piles.

Both structural steel and concrete spandrel beams were considered, with the latter being selected as they best suited architectural and fire-rating requirements. The 1220- by 305-mm (48- by 12-in.) beam rigidly connected to the large exterior columns and the small story height created a frame easily capable of resisting the longitudinal wind forces. The steel fabricator cut the 44- and 57-mm-diameter (#14 and #18) reinforcing bars and welded them to steel T sections holed for bolting at each end before delivering them to the precaster. The finished beams included a shear key and reinforcement for connection to the slabs.

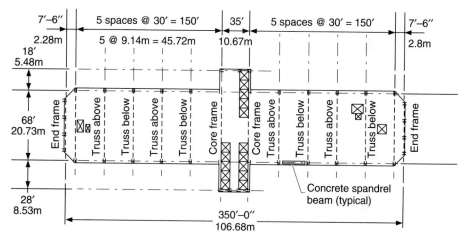

Fig. 4.33   Typical floor plan; Taj Mahal Hotel.

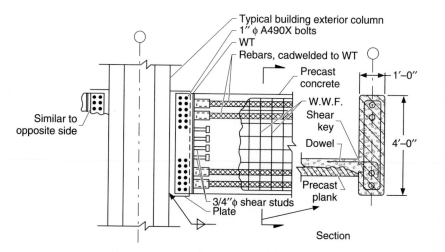

Fig. 4.34   Spandrel beam detail; Taj Mahal Hotel.

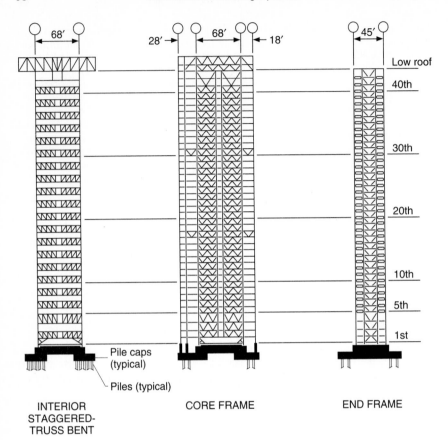

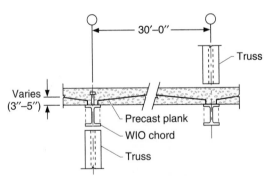

Fig. 4.35   Framework and typical bay section; Taj Mahal Hotel.

Tokyo Marine Building
Osaka, Japan

Architect	Kajima Design
Structural engineer	Kajima Design
Year of completion	1990
Height from street to roof	118 m (387 ft)
Number of stories	27
Number of levels below ground	3
Building use	Office, retail space, parking
Frame material	Steel
Typical floor live load	0.79 kPa (16.4 psf)
Basic wind velocity	35 m/sec (78 mph) at 10-m height
Maximum lateral deflection	Seismic control
Design fundamental period	3.31 sec longitudinal; 3.95 sec transverse
Design acceleration	4 mg peak, 5-yr return
Design damping	2%
Earthquake loading	Base shear coefficient 0.08
Type of structure	Moment resisting frame
Foundation conditions	Fine sand
Footing type	Cast-in-place concrete pile; 22-m (72-ft) length, 2.4-m (7.9-ft) diameter, with 4-m (13-ft) bell
Typical floor	
Story height	3.9 m (12 ft 9 in.)
Beam span	21 m (69 ft)
Beam depth	900 mm (36 in.)
Beam spacing	6.75 m (22 ft)
Material	Steel, grade SM 490, 483-MPa (70 ksi) tensile strength and below
Slab	155-mm (6.1-in.) lightweight concrete slab on corrugated deck
Columns	
Size at ground floor	500 by 500 mm (20 by 20 in.)
Spacing	10.8, 21 m (35, 69 ft)
Material	Steel, grade SM 490

The Tokyo Marine building is a 27-story office building located in the Osaka business park district being developed just east of Osaka Castle, Japan (Fig. 4.36). The building provides about 69,000 m^2 (743,000 ft^2) of area for offices, retail, and parking. Construction was completed in 1990. Architecturally the building was conceived to fit into the environment of the Osaka business park and to reflect the image of the client, Tokyo Marine. As the base of operations in western Japan for Tokyo Marine, the building was

Fig. 4.36   Tokyo Marine building, Osaka, Japan.

designed to have high-tech capabilities, to reflect its prestige appropriately by its exter-nal appearance, and also to be attractive to tenants as office space.

The building has a rectangular plan to fit into the site surrounded by two high-rise buildings on the longer sides. The exterior facade, exposing columns and beams outside the building, brings to mind the simple lines of traditional Japanese wood-frame details and gives a clear identity as well (Fig. 4.37). The lateral force resisting system of the

Fig. 4.37   Pilotis columns; Tokyo Marine building.

building consists of the framed columns interconnected with long-span beams [10.8, 21-m (70-ft) spans]. This allows for an open public area at the plaza level of the building and column-free space across the width of the building with a 2.7-m (9-ft) ceiling height on the office floors.

The building is framed in structural steel (Fig. 4.38). Each frame column consists of four vertical members joined by short [2.7-m (9-ft)-span] beams to create a three-dimensional structure. The combination of the short beams in between the framed column elements and the long-span beams creates unusual static characteristics: migration of column loads, the short-span beams being loaded more lightly in bending than the long-span beams, and an unusual failure-hinge mechanism for extreme seismic events.

In a medium-rise building, short-span beams yield at their supports under relatively low loading levels due to a concentration of bending stiffness. However, in this case the effect of axial deformations of the frame columns was the dominant mode of behavior. An incrementally increasing static analysis on an elastic-plastic model ("pushover" analysis) was carried out to obtain the skeleton curves for story shear. From this analysis the formation of plastic hinges at the supports of the long-span beams, with the short-span beams remaining elastic even at high earthquake levels, was notable.

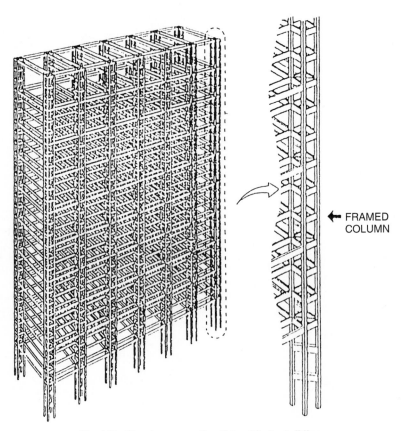

← FRAMED
COLUMN

Fig. 4.38 Framing perspective; Tokyo Marine building.

The structure was planned with an unusually long main span of 21 m (70 ft) across the full building width. In such cases the vertical component of an earthquake can have a significant effect on the beam stresses. This was investigated by modeling a typical bay as a two-dimensional frame. Dynamic analysis was carried out using the time history of four earthquake records, inputting their lateral and vertical components simultaneously. As a preliminary step, the modes of vibration of the structure were obtained, from which it was found that the fifth mode was the first mode in the vertical direction and involved axial tension and compression in the columns, with all the beams vibrating together. Only at higher modes did the beam vibration become more complex.

The analysis was carried out at two levels of earthquake input. The following were the results.

1. At 250-mm/sec (10-in./sec) peak input level, the long-span beams developed large bending moments, in particular high up in the building. It was verified that even including these stresses, the beams remained within the short-term allowable stresses according to the Japanese design criteria.

2. At 400-mm/sec (80-in./sec) peak input level, due to the effect of the vertical component of the earthquake, plastic hinges at the ends of long-span beams developed early on, but the structure remained elastic in its overall behavior and did not degrade its overall dynamic stiffness characteristics. The reason for this is that vertically the period of vibration is much shorter than horizontally, such that the structure is able to recover its elastic characteristics. From this result it was concluded that using horizontal earthquake time history records only for the main response analysis was appropriate.

Although the building is not of irregular shape, wind tunnel testing was carried out to investigate the effect of the exposed column frames on the surface roughness and effective frontal area of the building. It was also desirable to predict the vibration behavior of the building under wind loading, since its period is relatively long. Using the pressure coefficients from the wind tunnel test, the base shear from wind is about 91% that of seismic events. In terms of acceleration, it was predicted that although the upper levels would experience a peak acceleration of around 4 mg, this would not cause any discomfort.

It was considered that since the exposed frame was outside the main building glazing line, it would not be subject to the same intensity of fire as a normal frame, and therefore could be fire-protected to an appropriately lesser degree. An analysis was performed on the frame when subjected to flames discharging from inside the building. The results showed that the beams would be heated to 273°C and the columns to 281°C on the outside of their aluminum cladding. Since the critical temperature for steel in a fire may be taken as 350°C average (with a maximum of 450°C), the conclusion was drawn that no fire protection was needed at all for the external frame. Upon examination by the Building Center Fire Safety and Protection Committee, 10 mm (0.4 in.) of fire-resistant cladding material was finally agreed upon, which represented a substantial reduction in fireproofing material.

Kamogawa Grand Tower
Kamogawa, Japan

Architect	Kajima Design
Structural engineer	Kajima Design
Year of completion	1992
Height from street to roof	105 m (344 ft)
Number of stories	33
Number of levels below ground	1
Building use	Hotel and condominium
Frame material	Concrete
Typical floor live load	1.77 kPa (36.9 psf)
Basic wind velocity	35 m/sec (78 mph) at 10-m (33-ft) height
Design fundamental period	1.92 sec transverse; 1.55 sec longitudinal
Design damping	2%
Earthquake loading	Shear coefficient 0.085
Type of structure	Moment resisting frame with honeycomb damper wall
Foundation conditions	Fine sandy layer over clay layer over shale rock layer
Footing type	Cast-in-place concrete pile; 19-m (62-ft) length, 1.7-m (5.6-ft) diameter, with 2.8-m (9.2-ft) bell
Typical floor	
Story height	2.85 m (9 ft 4 in.)
Beam span	4.5 and 9 m (15 and 30 ft)
Beam depth	700 mm (27.5 in.)
Material	Reinforced concrete, normal weight
Slab	160-mm (6.3-in.) concrete
Columns	
Size at ground floor	900 by 900 mm (35.5 by 35.5 in.)
Material	Concrete, 23 to 41 MPa (3400 to 6000 psi)

The Kamogawa Grand Tower is composed of ductile moment resisting frames (Fig. 4.39). The high-rise reinforced concrete (HiRC) construction method developed by Kajima Corporation was used. It consists of pure reinforced concrete columns, girders, and floor slabs which are cast on site (Fig. 4.40). The typical floor plan is a stair-shaped plan along the exterior zones, which consists of regular square units 4.5 m (15 ft) on a side (Fig. 4.41). The standard span is this 4.5 m (15 ft) except at the central corridor, where the span is 9.0 m (30 ft) by skipping columns sustained by the cross girders. The entire structure is designed to be approximately symmetric along 45 and 135° orientations from the orthogonal so that earthquake resistance is balanced in all lateral directions.

Fig. 4.39 Kamogawa Grand Tower, Kamogawa, Japan.

On typical floors, steel plates with honeycomb-shaped openings are installed in the central corridor connecting to the cross girders (Fig. 4.42). Post columns extending from the midspan of upper- and lower-story girders are spliced at midstory using these damper plates, connected by high-strength bolts through gusset plates. Thus the story shear drift is concentrated in the damper plates. Sixteen units of damper plates and post

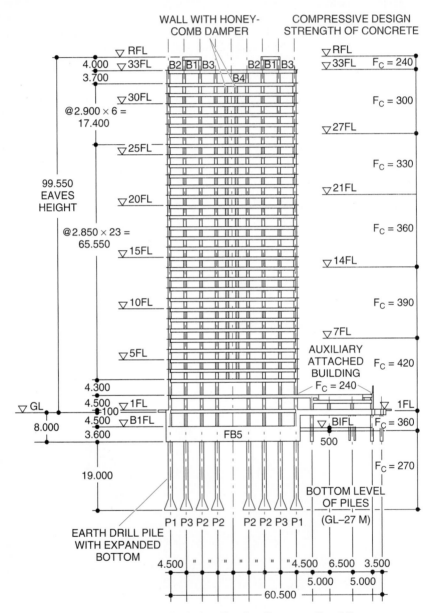

Fig. 4.40 Framing elevation in the *y* direction; Kamogawa Grand Tower.

columns are installed in each typical story. The seismic response of the building is reduced by the hysteresis damping effect due to the yielding of the steel plates.

The seismic design criteria for two levels of design earthquake were established as
follows:

1. *Severe earthquake:*   The stresses in all structural members must be less than the
   allowable values and the story drift must be less than 1/200.

2. *Worst earthquake:*   Even if the structural members exceed allowable limits, excessive large plastic deformation should not be caused and the story drift must be
   less than 1/100.

Referring to the preliminary earthquake response results in consideration of the hysteresis steel damper, the story shear coefficients are determined. The design wind shears
are about 56% of the seismic values. In order to secure the ultimate strength of the structural frame, it was established that the story shear capacity would be 1.5 times that of
the design earthquake shear forces. The ultimate bending and shear strength of the
columns was designed to be at least 1.25 times greater than that of the girder, so that
yielding to bending in the girders precedes yield in the columns; but at the tops of the
columns, in the top story and at the bottom of the first story, the bending yield in the
columns is considered.

From the earthquake responses of the structure to both severe and worst-case events
it was assured that the final design of the moment resisting frame structure was completely satisfactory with respect to the design criteria. Moreover, using the honeycomb
steel plates as hysteresis dampers, not only a well-balanced structure but also savings in
the volume of reinforced concrete may be realized.

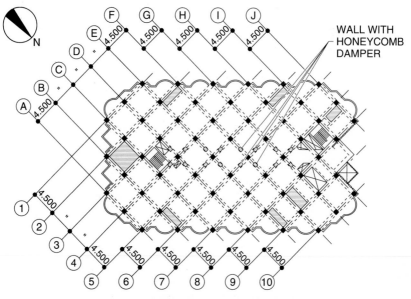

Fig. 4.41   Typical floor plan; Kamogawa Grand Tower.

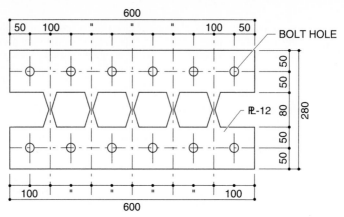

SHAPE OF HONEYCOMB DAMPER PLATE

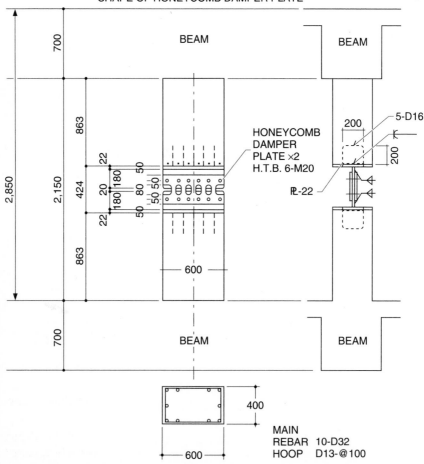

Fig. 4.42 Shape and installation of honeycomb damper plate; Kamogawa Grand Tower.

108

4.2 SHEAR WALL SYSTEMS

Shear walls have been the most common structural systems used in the past for stabilizing building structures against horizontal forces caused by wind or earthquakes. With the advent of reinforced concrete, shear wall systems have become widely used to stabilize efficiently even the tallest building structures. In the last 10 years, concrete technology has advanced to a point where concrete strengths of over 130 MPa (19,000 psi) are achievable in the field. This has led to the design of the proposed 610-m (2000-ft) Miglin-Beitler Tower in Chicago (which would become the world's tallest building), relying heavily on a shear wall system of very-high-strength concrete to resist horizontal forces.

A common shear wall system used for tall office buildings groups shear walls around service cores, elevator shafts, and stairwells to form a stiff box-type structure, such as for the Melbourne Central building in Australia (Chapter 3.2). In this example the need to enclose and fire-protect 21 passenger elevators, service elevators, two stairwells, lobbies, and service risers created the framework for a stiff concrete box-type shear wall system.

In contrast with office buildings, high-rise residential buildings have less demand for elevators, lobbies, and services, and hence do not usually have large stiff concrete shear wall boxes to resist horizontal forces. A more common system will incorporate a small box structure around a smaller number of elevators and stairwells, and include discrete shear walls between apartments.

In both shear wall systems noted, the walls are designed to cantilever from the foundation level. To design shear walls arranged around service cores, the bending, shear, and warping stresses due to wind or earthquake loads are combined with stresses due to gravity loads. Individual walls within the box system can then be designed as unit-length walls spanning either floor to floor or between return walls. Reinforcement is proportioned as follows:

1. Minimum shrinkage restraint reinforcement where the wall stresses are low, which can be for a substantial portion of the shear wall.

2. Tensile reinforcement for areas where tension stresses occur in walls when wind uplift stresses exceed gravity stresses.

3. Compressive reinforcement with confinement ties where high compressive forces require that walls be designed as columns. Individual shear walls, say at the edge of a tall building, are designed either as blade walls or as columns resisting shear and bending as required.

Multiple shear walls throughout a tall building may be coupled to provide additional frame action and hence increase overall building stiffness. Coupling can be realized by relatively shallow header or link beams within the ceiling cavity at each level or by means of one- or two-story-high shear coupling walls. By adding a coupling shear wall at a single level, reverse curvature is induced in the core above the coupling shear wall, significantly reducing lateral drift by increasing the overall building stiffness. As the increase in mass is minimal, there will be an increase in the building's natural frequency. This can be a desirable effect, in particular with respect to achieving an acceptable wind-induced acceleration response to ensure occupant comfort. Central core boxes can also be coupled via stiff beams or trusses, at discrete levels, to external shear walls or columns to achieve a similar and more pronounced effect than that noted. Thus the concrete shear wall becomes the central component in a core and outrigger system.

Many tall buildings undergo torsional loading due to nonalignment of the building shear center with the location of the horizontal load application. Such a situation occurs in the CitySpire Building (Chapter 4.3) due to the asymmetry of the location of the shear wall boxes. Torsional loading can also be induced in a building such as Bourke Place (Chapter

3.2) due to the periodic shedding of wind vortices alternately from each side of the structure, moving the instantaneous center of pressure out of line with the building's shear center.

Boxed shear wall systems provide an efficient means of resisting such torsion. Torsion is resisted by both warping and uniform shear. Particular care must be taken during computer modeling of boxed shear walls to reflect penetrations for elevator and stair doors. Calculation of inertias based on a reduced wall thickness, depending on the number of shear wall penetrations, is common. Boxed shear wall systems are very well suited to regular plan office buildings, as demonstrated in many of the project examples in this section. Construction advantages of reinforced concrete shear wall systems include the following:

1. Central-services core shear walls can be efficiently constructed using slip-form or jump-form techniques. In the case of 120 Collins Street, a $4\frac{1}{2}$-day cycle was achieved, ensuring that core wall construction was well off the critical path.

2. High-strength concrete has enabled wall thicknesses to be minimized, hence maximizing rentable floor space.

3. Technology exists to pump and place high-strength concrete at high elevations.

4. Fire rating for service and passenger elevator shafts is achieved by simply placing concrete of a determined thickness.

5. The need for complex bolted or site-welded steel connections is avoided.

6. Well-detailed reinforced concrete will develop about twice as much damping as structural steel. This is an advantage where acceleration serviceability is a critical limit state, or for ultimate limit state design in earthquake-prone areas.

Although these advantages make concrete shear wall systems a competitive construction method, the following must also be considered:

1. Shear walls formed around elevator and service risers require a concentration of openings at ground level where stresses are critical.

2. Torsional and flexural rigidity is affected significantly by the number and size of openings around the shear walls throughout the height of the building.

3. In 1 and 2 it is difficult to gauge the effect of openings precisely without undertaking time-consuming finite-element analysis.

4. Shear wall vertical movements will continue throughout the life of the building. Their impact on the integrity of the structure must be evaluated at the design stage.

5. Construction time is generally slower than for a steel-framed building.

6. The additional weight of the vertical concrete elements as compared to steel will induce a cost penalty for the foundations.

7. An increase in mass will cause a decrease in natural frequency and hence will most likely produce an adverse effect of the acceleration response depending on the frequency range of the building. But shear wall systems are usually stiff and cause a compensating increase in natural frequency.

8. There are problems associated with moving formwork systems, including the following:
   a. A significant time lag will occur between footing construction and wall construction because of the fabrication and erection on site of the moving formwork system.
   b. Time will be lost at levels where walls are terminated or decreased in thickness.
   c. Regular survey checks must be undertaken to ensure that the vertical and twist alignments of the shear walls are within tolerance.
   d. In general it is difficult to achieve a good finish from slip-form formwork systems, and hence rendering or some other type of finishing may be necessary.
   e. When walls are too thin [such as 150 mm (6 in.)] it is not unusual for friction between the forms and concrete to lift the concrete in slip-form construction, leading to cracks or gaping holes in the wall.

Project Descriptions

Metropolitan Tower
New York, N.Y., USA

Architect	Schuman, Lichtenstein, Claman and Efron with design input from Mackbowe/Denman/Werdiger
Structural engineer	Robert Rosenwasser Associates P.C.
Year of completion	1985
Height from street to roof	218 m (716 ft)
Number of stories	68
Number of levels below ground	2
Building use	Office to 18th floor; residential above
Frame material	Concrete
Typical floor live load	2.5 kPa (50 psf) office; 2 kPa (40 psf) residential
Basic wind velocity	47 m/sec (105 mph), 100-yr return
Maximum lateral deflection	$H/500$
Design fundamental period	5 and 4 sec horizontal; 2 sec torsion
Design acceleration	15 mg peak
Design damping	$1\frac{1}{4}\%$ serviceability; $2\frac{1}{2}\%$ ultimate
Earthquake loading	Not applicable
Type of structure	Coupled shear walls plus perimeter frames
Foundation conditions	Rock, 4-MPa (40-ton/ft^2) capacity
Footing type	Spread footings
Typical floor	
Story height	3.45 m (11 ft 4 in.) office; 2.95 m (9 ft 8 in.) residential
Beams	Span and spacing vary
Beam depth	508 mm (20 in.) at perimeter
Slab	216-mm (8.5-in.) flat slab
Material	Concrete, 42 to 28 MPa (6000 to 4000 psi)
Columns	Size and spacing vary
Material	Concrete, 58 to 39 MPa (8300 to 5600 psi)
Core	Coupled shear walls; thickness varies
Material	Concrete, 58 to 39 MPa (8300 to 5600 psi)

A rectangular tower would not work because of restrictions on the north-south-oriented site. This problem was solved with a triangular tower whose longest face is oriented northeast, with setbacks designed to conform to zoning regulations. The L-shaped commercial base is 18 stories, whereas the upper triangular condominium tower is 46 stories plus two stories for the mechanical and structural transition. The leading edge of the triangular tower facing north on 57th Street is continued for the entire 218-m (716-ft) height of the building, integrating the two basic forms. In this way the unique triangular tower maximizes one of its greatest assets—the views (Fig. 4.43).

Fig. 4.43 Metropolitan Tower, New York. (*Courtesy of Robert Rosenwasser Assoc.*)

The upper condominium tower contains 246 luxury apartments totaling 39,300 m² (423,000 ft²). The lower commercial base has 21,000 m² (225,000 ft²) of rental office space and 460 m² (5000 ft²) for retail rental. The total project amounts to 60,600 m² (653,000 ft²) and required approximately 23,000 m³ (30,000 yd³) of concrete and 3300 tonnes (3600 tons) of reinforcing steel. To keep an efficient column grid on the commercial floors, a double-height reinforced concrete mechanical floor was created at the nineteenth floor to allow the transfer of loads from the triangular plan of the building's upper tower to its L-shaped base (Fig. 4.44a). In effect this was a new foundation for the triangular tower, accomplished by using an extraordinary volume of concrete, an unusually dense mass of reinforcing steel, and beams up to 4 m (13 ft) deep. These transfer girders were cast in two stages, the bottom 600 to 900 mm (2 to 3 ft) being cast first to serve as support for the remainder of the concrete in the second placement.

The depth of the meandering shear wall (the main structural support of the triangular footprint) is about 21 m (70 ft) (Fig. 4.44b). Of the three available faces, the west face was a lot-line face, and therefore a place to accommodate the elevator shafts for the high-rise structure. It was recognized, and later verified in a wind tunnel test, that the structure would support larger wind forces acting perpendicular to the hypotenuse of the triangle.

Vortex shedding, which usually produces larger forces transverse to the wind direction, did not materialize for this structure because of its triangular footprint. Shear walls then migrate from the west lot line, meandering alongside the apartment lobby and corridors, to the hypotenuse side of the triangle, where additional columns were engaged via Vierendeel action of the spandrel beams. Other frame elements, 508-mm (20-in.)-deep spandrel beams along the periphery and 216-mm (8.5-in.) slabs at the interior of the structure, were needed to help counter large torsional loads since it was impossible to minimize torsional forces for all possible wind directions. This slender tower was somewhat stiffened by a wider base below the eighteenth floor. However, part of the shear wall and many of the columns had to be transferred utilizing deep concrete girders at this level. These deep girders were utilized, via outrigger action, to engage additional supports to help divert hold-down loads for the shear wall and to equalize the strain in the supports.

The flat slab floors are supported by a hybrid building frame of columns and shear walls, in part because of the developer's desire to leave the perimeter as column-free as possible. In the triangular tower, wind on the long side of the triangle governs the design, so the shear walls were placed at right angles to that face of the building, meandering along partition lines in a horseshoe shape to the opposite side of the tower and back to the long side of the triangle.

Several factors contributed to the decision to use concrete rather than steel. These included the easier modeling of shapes, the ability to make last-minute changes, and the knowledge that a larger mass reduces vibration and the perception of motion. The choice of concrete reflects the needs of the extremely tall slender structure. Sway of the building was an important concern. In high-rise buildings it may range from 1/500 to 1/600 of the building height in a 100-year wind (that is, the strongest wind that may be anticipated to occur in a 100-year period). When comparing buildings of structural steel and reinforced concrete having similar stiffnesses and movements, the perceived motion in the concrete building will be less because the larger mass of the concrete structure slows down its swaying motions, that is, the period is increased and the acceleration reduced.

In the Metropolitan Tower the typical slab floor thickness of 216 mm (8.5 in.) of stone concrete is important in achieving the mass of the building. Nevertheless, the building was designed with provisions to support the weight of a pendulum-type damper should it be needed. Using three accelerometers, field measurements were taken when the structure reached its fifty-fourth floor and, later on, at its sixty-sixth floor (at the last

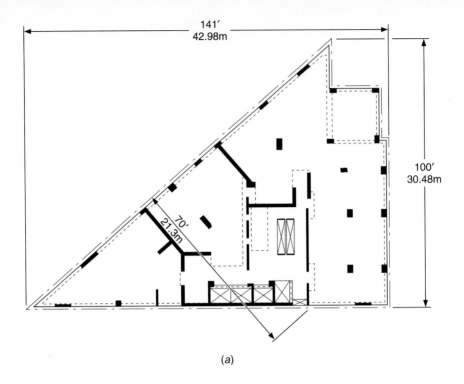

(a)

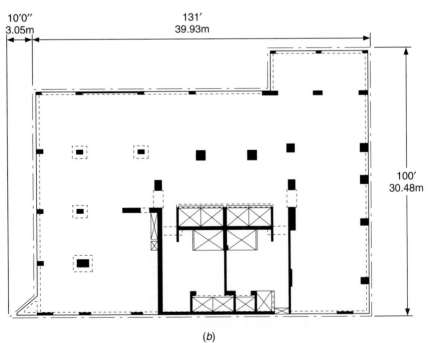

(b)

Fig. 4.44 Metropolitan Tower. (a) L-shaped base. (b) Meandering shear wall.

possible date, allowing time for a "go/no go" decision with regard to the installation of a damper), indicating that a damper was not needed. The extra cost to the owner resulted from providing a double design layout, with and without the damper. No materials, except those needed to support the damper's weight on the footings and columns, were actually expended in the structure. This structure can accommodate a future damper, if found necessary during its service life, with some minor modifications and rerouting of some mechanical pipes.

Slab formwork was cycled by the "preshoring" method commonly used in New York (Grossman, 1990). The first 18 stories, larger in floor area, were completed at the rate of about 4 to 5 days per story. In the triangular tower, two floors per week was typical progress, with columns and shear walls cast on Mondays and Thursdays and floors on Tuesdays and Fridays. Near the top of the tower, work speeded up to 2 days per story. The concrete frame was topped out on October 2, 1985.

Embassy Suites Hotel
New York, N.Y., USA

Architect	Fox and Fowle Architects
Structural engineer	DeSimone, Chaplin and Dobryn
Year of completion	1990
Height from street to roof	146.3 m (480 ft)
Number of stories	46
Number of levels below ground	1
Building use	Hotel
Frame material	Concrete above 8th floor; concrete-encased steel below
Typical floor live load	2 kPa (40 psf)
Basic wind velocity	36 m/sec (80 mph)
Maximum lateral deflection	$H/450$, $H/700$
Design fundamental period	5.4, 7.4 sec
Design acceleration	21 mg peak
Design damping	2% serviceability
Earthquake loading	$Z = 0.375$; $C = 0.030$ and 0.025; $K = 1.0$
Type of structure	Shear walls above 8th floor; encased-steel transfer trusses to steel supercolumns below
Foundation conditions	Rock, 4-MPa (40-ton/ft^2) capacity
Footing type	Concrete piers
Typical floor	
Story height	2.65 m (8 ft 8.5 in.)
Slab	200-mm (8-in.) flat plate, spanning 7.32 by 7.32 m (24 by 24 ft)
Columns	4 supercolumns built up from five 200-mm (8-in.) plates on 14.63- by 39.63-m (48- by 130-ft) grid
Core	Shear walls, 300 to 450 mm (12 to 18 in.) thick at ground floor
Material	56-MPa (8000-psi) concrete

1568 Broadway is the site of the Embassy Suites Hotel in the Times Square district of New York City (Fig. 4.45). It is built over the historic Palace Theatre, a landmark dating back to 1919. Because of the theater's landmark status, New York City would not permit any disturbance to the theater by the new hotel. It was therefore necessary to support this 46-story, 146-m (480-ft)-tall building by building a "bridge" over the theater (Fig. 4.46).

The transfer was accomplished with a hybrid composite steel and concrete structure consisting of two 40-m (130-ft)-long composite trusses and steel cross trusses. Four supercolumns, two on either side of the theater, come down to ground to support the structure. These columns were built up out to thick grade 350-MPa (50-ksi) steel plates and weigh up to 6000 kg/m (4000 lb/ft). The truss members were designed to be light enough to permit erection on an extremely difficult site. To give them the necessary stiffness, the entire trusses were encased in concrete. The ballrooms, kitchen, and mechanical spaces are located between the 14.9-m (49-ft)-high trusses. The system is efficient and economical and solved the problems associated with constructing over a landmark.

The hotel superstructure is a reinforced concrete flat-plate system with a 8.5- by 8.5-m (28- by 28-ft) column grid and was built on a 2-day cycle. Wind is resisted by shear walls as well as moment frame action of slab strips and columns. The total weight of the reinforcing steel used for the concrete tower was only 36.7 kg/m² (7.5 psf).

Fig. 4.45   Embassy Suites Hotel, New York.

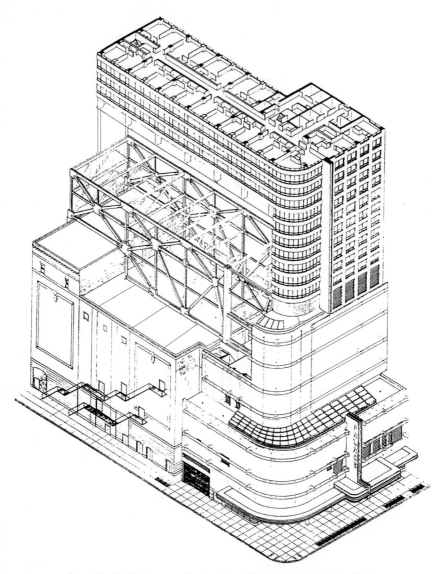

Fig. 4.46   "Bridge" supporting hotel over theater. Embassy Suites Hotel.

Singapore Treasury Building
Singapore

Architect	Hugh Stubbins and Associates
Structural engineer	LeMessurier Consultants with Ove Arup and Partners
Year of completion	1986
Height from street to roof	234 m (768 ft)
Number of stories	52
Number of levels below ground	5
Building use	Office
Frame material	Concrete core, steel floor beams
Typical floor live load	2.5 kPa (50 psf) 30th floor and above; 3.0 kPa (60 psf) below 30th floor
Basic wind velocity	38 m/sec (85 mph)
Design fundamental period	5.6 sec
Design acceleration	Not estimated
Design damping	Approx 2% serviceability
Earthquake loading	Not applicable
Type of structure	Steel floor beams cantilevered off cylindrical concrete core wall
Foundation conditions	Clay over rock
Footing type	6 8-m (26-ft 3-in.)-diameter caissons, 35 m (115 ft) long, under a 2.9-m (9-ft 6-in.)-thick mat
Typical floor	
Story height	4.25 m (13 ft 11 in.)
Beams	Cantilever 11.58 m (38 ft), spacing 4.9 m (16.42 ft) at core
Beam depth	1470 mm (58 in.), facade truss 1260 mm (50 in.) deep, continuous
Slab	80 mm (3.25 in.) on 77-mm (3-in.) steel deck
Columns	Only erection columns embedded in core wall
Core	Reinforced concrete cylinder, 22.95-m (75-ft) I.D., 1.65 to 1 m (65 to 39 in.) thick
Material	Concrete cube, 40 to 30 MPa (4500 to 3400 psi)

This cylindrical 48.4-m (159-ft)-diameter mixed construction office tower, located in the center of Singapore, has an area of more than 132,000 m^2 (1.42 million ft^2) (Fig. 4.47). Although the Singapore wind climate is relatively benign, avoidance of resonant vibration caused by wind-induced vortex shedding controlled the required lateral stiffness of the tower. This required setting the first vibration mode period at no more than 5.6 sec.

The architect and owner wanted to have little or no visible structure obstructing the 360° panoramic sweep of the windows at each floor. The simple yet elegant structural solution was to cantilever every floor from an inner cylindrical wall enclosing the elevator and service core. This required radial beams which cantilever 11.6 m (38 ft) from the 24.95-m (81.8-ft)-outside-diameter reinforced concrete core wall. Each cantilever girder is welded to a steel erection column embedded in the core wall (Fig. 4.48). The cantilevers on successive floors are connected at their outer ends by 25- by 100-mm (1- by 4-in.) steel ties, hidden in the curtain wall, which reduce relative vertical deflections

Fig. 4.47   Singapore Treasury Building, Singapore. (*Courtesy of The Stubbins Association.*)

of adjacent floors. A stiff continuous perimeter ring truss at each floor minimizes relative deflections of adjacent cantilevers on the same floor produced by any uneven live loading. This truss plus the vertical ties also provide some redundancy in the unlikely event of a cantilever failing.

All gravity load and all the wind loads are resisted by the concrete core wall. For strength alone, the core wall would have been a constant thickness almost to ground level, but in order to meet the building period limitation, it was necessary to thicken the wall from its typical 1.0-m (3.3-ft) dimension to 1.2 m (4 ft) and then 1.65 m (5.4 ft) below the sixteenth floor. A concrete core wall was selected in lieu of an all-steel diago-

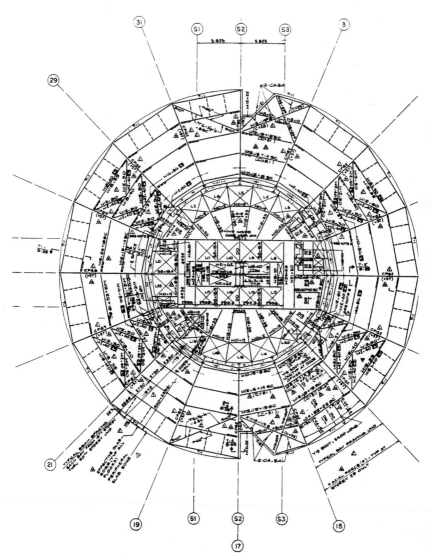

Fig. 4.48   Typical floor plan; Singapore Treasury Building.

nally braced "wall" for reasons of economy. The core is spanned by two plate girders. The core wall has four doorway openings on each floor. The headers over these openings consist of rigid steel Vierendeel girders, which allow duct work to pass through (Fig. 4.49).

Structural steel floor framing was used to facilitate a modular electrified underfloor steel deck, including trench headers, and to make the long cantilevers quite stiff. Typical live-load deflection at the end of the cantilever was less than 25 mm (1 in.). Girders were cambered to counteract dead-load deflection. Web openings were provided in the cantilevers for ducts and pipes. To verify the design and fabrication quality and reassure the owner that deflections would not be excessive, a full-size prototype cantilever girder welded to a two-story steel column was tested at the steel fabricator's laboratory in Japan. The test was quite successful and verified the accuracy of the structural analysis within a few percent. This was the first significant steel-framed building to be built in Singapore, so the test was also helpful in providing assurance to the building officials of the competence of the design and steel construction team.

Because of the somewhat unusual structural framing system, the concrete core wall was designed conservatively to resist possible, although very unlikely, pattern live loadings in which several consecutive floors had live loads in certain quadrants and no live load in others. The result of such loading patterns was to induce through-thickness bending stresses in the wall due to these asymmetrical forces. The core wall was analyzed using detailed finite-element analyses, and reinforcing steel was provided to resist the in-plane and through-thickness forces and bending moments due to gravity loads with and without wind loads.

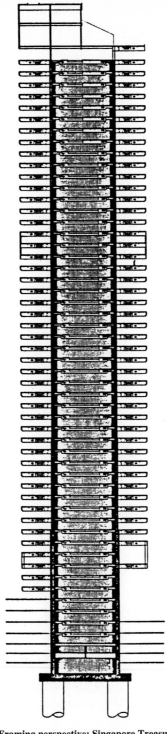

Fig. 4.49   Framing perspective; Singapore Treasury Building.

123

77 West Wacker Drive
Chicago, Illinois, USA

Architect	Richard Bofill/DeStefano and Goettsch
Structural engineer	Cohen-Barreto-Marchertas, Inc.
Year of completion	1992
Height from street to roof	203.6 m (668 ft)
Number of stories	50
Number of levels below ground	2
Building use	Office
Frame material	Concrete core, steel perimeter
Typical floor live load	2.5 kPa (50 psf)
Basic wind velocity	Chicago building code
Maximum lateral deflection	Less than $H/500$
Design fundamental period	6.67, 5.88 sec horizontal; 6.67 sec torsion
Design acceleration	29 mg peak
Design damping	2% serviceability
Earthquake loading	Not applicable
Type of structure	Concrete shear core, perimeter steel frames
Foundation conditions	Hardpan, 1700-kPa (40,000-psf) capacity
Footing type	21-m (70-ft)-deep caissons, 900- to 3000-mm (3- to 10-ft) shaft diameter belled to 1370 to 7000 mm (4.5 to 23 ft)
Typical floor	
Story height	3.96 m (13 ft 9 in.)
Beam span	13.72 m (45 ft)
Beam depth	533 mm (21 in.)
Beam spacing	3.43 m (11 ft 3 in.)
Material	Steel
Slab	140-mm (5.5-in.) concrete on metal deck
Columns	
Size at ground floor	W350 × 1088 (W14 × 730) plated
Column spacing	3 m (10 ft) min, 13.7 m (45 ft) max
Material	Steel, F_y = 350 MPa (50 ksi)
Core	Central concrete shear core
Wall thickness at ground floor	559 and 355 mm (22 and 14 in.)
Material	Concrete, 52 to 35 MPa (7500 to 5000 psi)

This 50-story 96,600-m^2 (1,040,000-ft^2) office tower is located at the southwest corner of Wacker Drive and Clark Street (Fig. 4.50). It is a classically styled addition to the Chicago skyline on North Wacker Drive, which is graced by several outstanding architectural and structural originals.

Fig. 4.50   77 West Wacker Drive, Chicago, Illinois.

The building, which is rectangular in shape, 50.29 by 42.67 m (165 by 140 ft) with 4.57-m (15-ft) reentrant angles at the four corners, is the first high-rise tower designed by the Spanish architect, Ricardo Bofill. It was designed in collaboration with the Chicago architectural firm of DeStefano and Partners.

The framing system is a central concrete core surrounded by a structural steel frame with a composite floor deck (Fig. 4.51). The core, which is extremely slender [15.55 by 27.45 m (51 by 90 ft) with a height-to-width ratio greater than 13:1] incorporates all the mechanical, electrical, and vertical transportation amenities. The column-free floor spans allow for a very flexible 13.72-m (45-ft)-wide tenant space.

Another outstanding feature in the building is its magnificent entrance lobby, which extends from the ground to the fifth floor, with a completely unobstructed space of 50.29 by 13.72 m (165 by 45 ft), 13.72 m (45 ft) high.

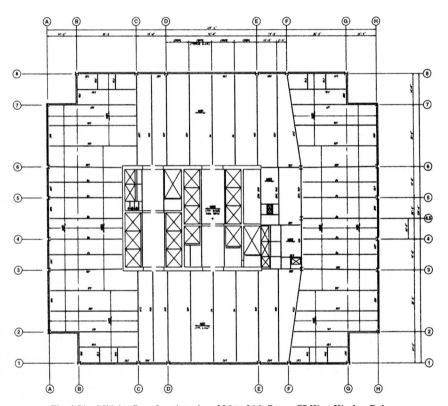

Fig. 4.51   Midrise floor framing plan, 23d to 36th floors; 77 West Wacker Drive.

Casselden Place
Melbourne, Australia

Architect	Australian Construction Services with Hassell Architects
Structural engineer	Connell Wagner
Year of completion	1992
Height from street to roof	160 m (525 ft)
Number of stories	43
Number of levels below ground	3
Building use	Office
Frame material	Concrete core, steel frame
Typical floor live load	4 kPa (80 psf)
Basic wind velocity	41 m/sec (92 mph), 50-yr return
Maximum lateral deflection	150 mm (6 in.), 1000-yr return
Design fundamental period	3.45, 5.00 sec
Design acceleration	4.5 mg rms, 5-yr return
Design damping	1% serviceability; 5% ultimate
Earthquake loading	Not applicable
Type of structure	Core for all lateral load
Foundation conditions	Siltstone, 2-MPa (20-ton/ft^2) capacity
Footing type	Pad footings
Typical floor	
Story height	3.75 m (12 ft 4 in.)
Beam span	12 m (39 ft 4 in.)
Beam depth	610 mm (24 in.)
Beam spacing	3 m (9.83 ft)
Slab	130 mm (5 in.) on metal deck
Columns	
Size at ground floor	950-mm (37-in.)-diameter composite concrete-filled steel tubes
Material	Concrete, 70 MPa (10,000 psi)
Core	Concrete shear walls, 500 and 200 mm (22 and 8 in.) thick at ground floor
Material	Concrete, 70 MPa (10,000 psi)

This building is interesting for several reasons:

1. Construction over Melbourne underground rail loop
2. Use of high-strength concrete
3. Use of composite concrete-filled steel-tube columns

The construction of Casselden Place (Fig. 4.52) over the Melbourne underground rail loop necessitated two unusual design features. (1) The removal of rock for the three-

Fig. 4.52 Casselden Place, Melbourne, Australia.

story basement relaxed the overburden pressure on the tunnels. (2) To prevent heaving of the tunnels, 26 30-tonne (33-ton) vertical anchors were installed to tie the tunnels down. In the areas where only light loads were reimposed, these anchors are permanent, but where heavy loads are imposed by the new structure, temporary anchors only were used. In addition, piling was used in some areas to provide load transfer to below the level of the tunnels in the event of ground movement.

The most interesting part of the construction is the columns construction. This method is the first of its type in Australia, with only a small number of buildings known to be constructed using similar methods anywhere in the world. The tube columns are erected in two-story lifts, with the bare steel able to support up to six stories of construction. Concrete is pumped into the base of the tube, and up as many as six stories at a time. No vibration of the concrete is required. Connell Wagner has developed design methods for this type of column, including the use of thin-walled tubes. No codified method for the design of thin-walled concrete-filled tubes is available anywhere in the world.

This form of construction provides a column for a steel-framing system at a cost equal to that of a reinforced concrete column. The cost of the columns has been a major stumbling block in the economies of steel-framed buildings, with the penalty for using all-steel columns on a building such as this as high as 3% of the total building value— millions of dollars on projects of this size. This solution benefits from the economy of concrete, with the simple concrete placement method giving the system constructability that is equivalent to that of a full steel column.

The core and columns on the project use concrete of up to 70 MPa (10,000 psi). The columns are considered to be an ideal way of using high-strength concrete of good curing ability, which is being placed inside a tube. The tube confines the concrete, enhancing the ductility of the high-strength materials.

Twin 21
Osaka, Japan

Architect	Nikken Sekkei Ltd.
Structural engineer	Nikken Sekkei Ltd.
Year of completion	1986
Height from street to roof	157 m (515 ft)
Number of stories	38
Number of levels below ground	1
Building use	Office, shops, showrooms
Frame material	Steel core and perimeter on upper floors; concrete core and concrete-encased steel perimeter on lower floors
Typical floor live load	3 kPa (60 psf)
Basic wind velocity	35 m/sec (78 mph)
Maximum lateral deflection	400 mm (16 in.)
Design fundamental period	3.9, 4.0 sec
Design velocity	250 mm/sec (10 in./sec) for medium earthquakes; 500 mm/sec (20 in./sec) for maximum-level earthquakes
Design damping	2%
Earthquake loading	$C = 0.10$
Type of structure	Primarily perimeter rigid moment frames
Foundation conditions	Clay
Footing type	18-m (59-ft)-long, 1.5- to 2-m (5- to 6.5-ft) shaft-diameter belled concrete piles
Typical floor	
Story height	3.75 m (12 ft 4 in.)
Beam span	13.7 m (45 ft)
Beam depth	820 mm (32 in.)
Beam spacing	3.2 m (10 ft 6 in.)
Material	Steel, grade SS 400 and SM 490
Slab	165-mm (6.5-in.) concrete on metal deck
Columns	
Size at ground floor	1400 mm (55 in.)
Spacing	6.4 and 12.8 m (21 and 42 ft)
Material	Reinforced concrete and structural steel
Core	Reinforced concrete lower levels; steel upper levels
Thickness at ground floor	700 to 900 mm (27 to 35 in.)

Twin 21 comprises two identical 38-story office towers with shops and showrooms on the lower floors (Fig. 4.53). The perimeter frames above the sixth floor have columns spaced at 3.2-m (10-ft 6-in.) intervals, connected by the floor slabs to the steel-framed core. This structure is efficient in resisting horizontal and torsional deformations due to earthquakes and wind.

Below the sixth floor the building structure consists of steel frames encased in reinforced concrete, and rigidity is provided by reinforced concrete shear walls around the core. The majority of the horizontal force is borne by these shear walls (Fig. 4.54).

Had the tower building columns been continued down through the low-rise section at 3.2-m (10-ft 6-in.) centers, space utilization would have been adversely affected. Hence the 3.2-m (10-ft 6-in.) spans are increased to 12.4-m (40-ft 8-in.) spans by one-story-high concrete-encased steel transfer beams at the fifth-floor level, thereby providing for shops and showrooms in the lower floors of the building.

The wind load response due to the twin towers being in close proximity was checked using wind tunnel testing and the results were reflected in the design.

The atrium of the low-rise part is surrounded by the low-rise parts of the two towers and the gallery building (four stories with an L-shaped floor plan). It is composed of a large space [about 47 by 47 m (156 by 156 ft)] and is covered by a large steel-pipe space truss roof structure.

There are large forces on the roof due to the uplift of the wind blowing between the twin towers and the down wash off the buildings. These factors were evaluated by wind tunnel testing.

The atrium roof trusses are supported on slide bearings, which can absorb horizontal deformations of the high-rise part during an earthquake. Stoppers are provided to prevent uplift under upward wind loading.

Fig. 4.53   Twin 21, Osaka, Japan.

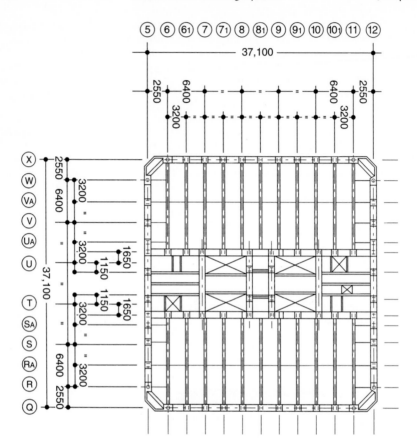

Fig. 4.54   **Typical structural floor plan; Twin 21.**

Majestic Building
Wellington, New Zealand

Architect	Manning and Associates
Structural engineer	Wass Buller and Associates
Year of completion	1991
Height from street to roof	116 m (380 ft)
Number of stories	29
Number of levels below ground	3
Building use	Office
Frame material	Concrete
Typical floor live load	3.5 kPa (70 psf)
Basic wind velocity	50 m/sec (112 mph)
Design fundamental period	2.9 sec
Design acceleration	10 mg peak, 1-yr return period
Design damping	1% serviceability (wind), 5% EQ
Earthquake loading	$C_d = 0.0432$
Type of structure	Core and perimeter frame
Foundation conditions	Weathered rock over rock
Footing type	Pads and 1.8-m (6-ft)-diameter bored piles
Typical floor	
Story height	3.7 m (14 ft 2 in.)
Beam span	12 m (39 ft 4 in.)
Beam depth	750 mm (29.5 in.)
Beam spacing	10 m (32 ft 10 in.)
Slab	365 mm (14 in.) Dycore
Columns	
Size at ground level	1400-mm (55-in.)-diameter
Spacing	10 m (32 ft 10 in.)
Material	Concrete, 50 MPa (7100 psi)
Core	
Thickness at ground level	400 and 600 mm (16 and 24 in.)
Material	Concrete, 50 MPa (7100 psi) max

The Majestic Building (Fig. 4.55) comprises 32 levels totaling 42,000 m² (452,000 ft²), including four levels of parking garage, extensive retail, arcade, and public plaza areas, a fitness center with a 33- by 4.5-m (110- by 15-ft) swimming pool, a crèche, an art gallery, and approximately 24,000 m² (258,000 ft²) of office space.

Wind engineering played a major part in determining the building shape, podium features, and structure of the building. Three separate wind tunnel studies were undertaken to investigate environmental wind effects, cladding pressures, as well as overturning moments and acceleration levels. Following completion, further studies of the structure were carried out using a mechanical vibrator and also recording wind displacements using sensitive accelerometers.

The building is located in the most active seismic zone of New Zealand, with known fault lines running through the central business district of Wellington. The first floor of the tower is 12 m (40 ft) above street level and the column spacing around the perimeter is 10 m (33 ft). These features were critical to create a spacious lobby and entrance to the building; however, such features in seismic zones require special design to prevent the occurrence of a "soft story." For these reasons a "ductile hybrid structure" was chosen as the lateral load resisting system. The concrete core walls and the perimeter frame work together and were designed using capacity design methods to be fully ductile. Foundations and lower levels were designed to resist the overstrength capacity forces from the superstructure.

The unique floor system comprises pretensioned hollow core planks 1200 mm (4 ft) wide and 300 mm (12 in.) deep, spaced at 2400-mm (8-ft) centers. A thin metal tray was placed between the hollow core planks, and 65 mm (2.5-in.) of in-situ concrete was placed over the whole floor. The floor is only 115 mm (4.5 in.) deep in parts, which allows for efficient duct layouts. It weighs only 3.6 kPa (75 psf) and can support in excess of 3.5 kPa (73 psf) over 12.5-m (41-ft) spans (Fig. 4.56).

Fig. 4.55   Majestic Building, Wellington, New Zealand. (*Courtesy of Wass Buller and Associates;
Photo by O'Neill's.*)

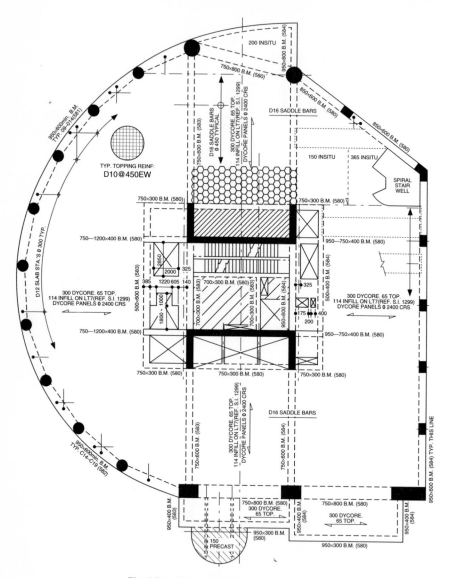

Fig. 4.56   Tower structure; Majestic Building.

Telecom Corporate Building
Melbourne, Australia

Architect	Perrott Lyon Mathieson
Structural engineer	Connell Wagner
Year of completion	1992
Height from street to roof	192 m (630 ft)
Number of stories	47
Number of levels below ground	3
Building use	Offices
Frame material	Concrete
Typical floor live load	4 kPa (80 psf)
Basic wind velocity	41 m/sec (92 mph), 50-yr return
Maximum lateral deflection	123 mm (5 in.) at 25 mm (1 in.), 100-yr return
Design fundamental period	4.5 sec
Design acceleration	4.4 mg rms, 5-yr return
Design damping	1% serviceability; 5% ultimate
Earthquake loading	Not applicable
Type of structure	Concrete core and perimeter frame tube in tube
Foundation conditions	Siltstone, 2000-kPa (20-ton/ft^2) capacity
Footing type	Pads to columns, raft to core
Typical floor	
Story height	3.85 m (12 ft 8 in.)
Beam span	12 m (39 ft 4 in.)
Beam depth	440 mm (16 in.)
Beam spacing	5 m (16 ft 5 in.)
Material	Partially prestressed concrete
Slabs	125-mm (5-in.) reinforced concrete
Columns	
Size to ground floor	1000 by 1200 mm (39 by 47 in.)
Spacing	8.1 or 9 m (26 ft 7 in. or 29 ft 6 in.)
Material	Concrete, 60 MPa (8500 psi)
Core	Shear walls
Thickness at ground floor	500 and 200 mm (20 and 8 in.)
Material	Concrete, 60 MPa (8500 psi) max

This all-concrete building achieved impressive construction times (Fig. 4.57). The entire 50-level concrete core was completed in 14 months, using a jump-form system. Typical cycle times for the core averaged $4\frac{1}{2}$ days per floor.

The tower floor band beams, typically 400 mm (16 in.) deep, are notched to 275 mm (11 in.) thick at the core to allow the major mechanical ring duct to encroach into the structural depth, thereby reducing the floor-to-floor height.

The band beams were designed as partially prestressed and are offset from the columns. A typical beam has three tendons. Two tendons, each with four 15.2-mm (0.6-in.)-diameter strands, were stressed from the external end of the beam. A single tendon

Fig. 4.57   Telecom Corporate Building, Melbourne, Australia. (*Photo by Squire Photographics.*)

with three 15.2-mm (0.6-in.) strands was tensioned from the opposite end. The bands were top-stressed. Grinding in back of the surface of the anchorage pockets was not necessary because access flooring is being provided throughout the tower.

One hundred percent of the prestress force was applied to each tendon when the concrete had reached a strength of 22 MPa (3100 psi). Using a high early-strength concrete mix, this was achieved on the second day after the pour. This, together with the use of two sets of table forms, allowed floor-to-floor cycles of three days to be achieved.

The tendons arrive on site prefabricated with strands already threaded into the ducts. The connection to the core is simply and positively affected by the use of 600-mm (24-in.)-long 20-mm (0.75-in.) bars which wrap around the vertical reinforcement in the core wall. The perimeter spandrel beams are 775 mm (30.5 in.) deep by 350 mm (14 in.) wide, spanning up to 9 m (30 ft). Reinforcement cages for these beams were fabricated on construction decks on the podium roof and craned directly into position. Loose bars were added at column locations to provide continuity.

The main entrance to the building is a dramatic three-story-high entry atrium. The perimeter of this atrium is glass on exposed architectural steelwork fabricated from 250-by 250-mm (10- by 10-in.)-square hollow sections. This steelwork is hung from a 2200-mm by 950-mm (86- by 37-in.) posttensioned cantilever ring beam at level 3, giving the impression of a glass cube suspended in midair. The ring beam is clad with 200-mm (8-in.)-thick polished precast panels used as formwork.

The entry space is further enhanced by the termination of one of the tower columns above the lobby level. The column load is 24,000 kN (2640 tons). This is achieved using stage-stressed 3950- by 1000-mm (155- by 39-in.) posttensioned beams, each spanning 18 m (59 ft) in a cruciform layout. The beams have eight and six tendons, respectively, with 19 12.5-mm (0.5-in.)-diameter strands in each tendon. The beams are stressed in three stages as load from the tower is progressively applied, achieving essentially flat beams throughout the construction phase.

4.3 CORE AND OUTRIGGER SYSTEMS

While outriggers have only been incorporated into high-rise buildings within the last 25 years, the outrigger as a structural element has a much longer history. The great sailing ships of the past and present have used outriggers to help resist the wind forces in their sails, making the adoption of tall and slender masts possible. In high-rise buildings the core can be related to the mast of the ship, with the outrigger acting like the spreaders and the exterior columns like the stays or shrouds. The typical organization of a core and outrigger system is pictured in Fig. 4.58. Just as in sailing ships, these outriggers serve to reduce the overturning moment in the core that would otherwise act as a pure cantilever, and to transfer the reduced moment to columns outside the core by way of a tension-compression couple, which takes advantage of the increased moment arm between these columns. In addition to reducing the size of the mast, the presence of outriggers also serves to reduce the critical connection where the mast is stepped to the keel beam. In high-rise buildings this same benefit is realized by a reduction of the base core overturning moments and the associated reduction in potential core uplift forces. The same overturning moment which is taken through a couple between the windward stay and the mast to the pretensioned ties in sailing ships, is transferred to gravity-loaded precompressed columns in the high-rise building.

The structural elegance and efficiency of outriggers are well rooted in history. The outriggers have also become key elements in the efficient and economic design of high-rise buildings.

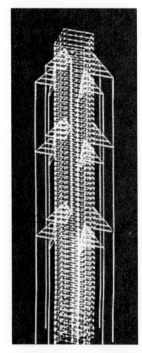

Fig. 4.58   Three-dimensional model after optimization; One Liberty Place, Philadelphia, Pennsylvania.

1 Why Outriggers?

Modern high-rise buildings frequently incorporate central elevator cores along with generous column-free floor space between the core and the exterior support columns. While this results in greater functional efficiency, it also effectively disconnects the two major structural elements available to resist the critical overturning forces present in a high-rise building. This uncoupling of the interior core and the perimeter frame reduces the overall resistance of the structure to the overturning forces to the sum of the independent resistances of the individual elements. The incorporation of outriggers in this same system couples these two components and enhances the system's ability to resist overturning forces dramatically.

For buildings of up to 35 to 40 stories, reinforced concrete shear wall or steel-braced cores have been effectively utilized as the sole lateral load resisting system. These systems are very effective in resisting the forces and associated deformations due to shear racking since their resistance varies approximately linearly with the building height. However, the resistance that core systems alone provide to the overturning component of drift decreases approximately with the cube of the height, so that such core systems become progressively more inefficient as the height of the building increases. In addition to stiffness limitations, a core system alone can also generate excessive uplift forces in the core structure along with prohibitively high overturning forces in the building's foundation system. With the system's inability to take advantage of the overall building depth, designing for the resulting uplift forces can be problematic.

In reinforced concrete cores, excessive or impractical wall elements where large net tension forces exist can negate the inherent efficiency of concrete in compression resistance. In steel cores, large and costly field-bolted or -welded tension splices greatly reduce steel efficiency and the ease of fabrication and erection.

In the foundation system, these uplift forces can lead to the need for the following:

- The addition of expensive and labor-intensive rock anchors to an otherwise "simple" foundation alternative such as spread footings.
- Greatly enlarged mat dimensions and depths solely to resist overturning forces.
- Time-consuming and costly rock sockets for caisson systems along with the need to develop reinforcement throughout the complete caisson depth.
- Expensive and intensive field-work connections at the interface between core and foundation. These connections can become particularly troublesome when one considers the difference in construction tolerances between foundation and core structure.
- The elimination from consideration of foundation systems which might have been considerably less expensive, such as piles, solely for their inability to resist significant uplift.

2 Outrigger Benefits

For many buildings, the answer to the problems and restrictions of core-only or tubular structures is the incorporation of one or more levels of outriggers. Typical outrigger organization consists of linking the core of a high-rise building to the exterior columns on one or more building faces with truss or wall elements (Fig. 4.59). The outrigger systems may be formed in any combination of steel, concrete, or composite construction. When properly and efficiently utilized, outriggers can provide the following structural and functional benefits to a building's overall design:

- Core overturning moments and their associated induced deformation can be reduced through the "reverse" moment applied to the core at each outrigger intersection (Fig. 4.60). This applies to the core at each outrigger intersection. This moment is created by the force couple in the exterior columns to which the outriggers connect. It can potentially increase the effective depth of the structural system from the core only to almost the complete building.

- Significant reduction and possibly the complete elimination of uplift and net tension forces throughout the columns and the foundation system.

- The exterior column spacing is not driven by structural considerations and can easily mesh with aesthetic and functional considerations.

- Exterior framing can consist of "simple" beam and column framing without the need for rigid-frame-type connections, resulting in economies.

- For rectangular buildings, outriggers can engage the middle columns on the long faces of the building under the application of wind loads in the more critical direc-

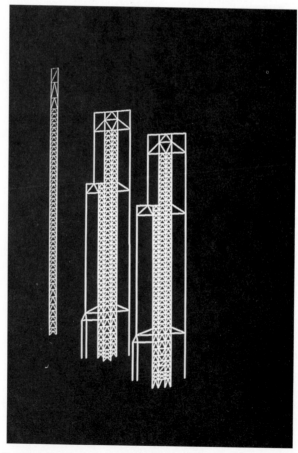

Fig. 4.59   Bond Building, Sydney, Australia. Tower bracing, east-west lines, looking north.

tion. In core-alone and tubular systems, these columns which carry significant gravity load are either not incorporated or underutilized. In some cases, outrigger systems can efficiently incorporate almost every gravity column into the lateral load resisting system, leading to significant economies.

3 Outrigger Drawbacks

The most significant drawback with the use of outrigger systems is their potential interference with occupiable and rentable space. This obstacle can be minimized or in some cases eliminated by incorporation of any of the following approaches:

- Locating outriggers in mechanical and interstitial levels
- Locating outriggers in the natural sloping lines of the building profile
- Incorporating multilevel single diagonal outriggers to minimize the member's interference on any single level
- Skewing and offsetting outriggers in order to mesh with the functional layout of the floor space

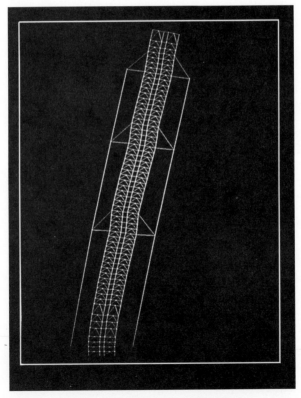

Fig. 4.60   1650 Market Street, Philadelphia.

Another potential drawback is the impact the outrigger installation can have on the erection process. As a typical building erection proceeds, the repetitive nature of the structural framing and the reduction in member sizes generally result in a learning curve which can speed the process along. The incorporation of an outrigger at intermediate or upper levels can, if not approached properly, have a negative impact on the erection process. Several steps can be taken to minimize this possibility.

- Provide clear and concise erection guidelines in the contract documents so that the erector can anticipate the constraints and limitations that the installation will impose.

- If possible, avoid outrigger locations or design constraints that will require "backtracking" in the construction process to install or connect the outrigger. The incorporation of intermediate outriggers in concrete construction or large variations in dead-load column stresses between the core and the exterior can in some cases result in the need to "backtrack." Such a need can be minimized if issues such as creep and differential shortening are carefully studied during the design process to minimize their impact.

- Avoid adding additional outrigger levels for borderline force or deflection control. Outriggers provide diminishing returns for each additional level added. Incorporate outriggers in less optimal numbers or locations when doing so will have a significant positive impact on the overall construction costs.

PROJECT DESCRIPTIONS

CitySpire
New York, N.Y., USA

Architect	Murphy Jahn
Structural engineer	Robert Rosenwasser Associates
Year of completion	1987
Height from street to roof	248 m (814 ft)
Number of stories	75
Number of levels below ground	2
Building use	Office and residential
Frame material	Concrete
Typical floor live load	2.5 and 2 kPa (50 and 40 psf)
Basic wind velocity	47 m/sec (105 mph), 100-yr return
Maximum lateral deflection	$H/500$
Design fundamental period	5.5, 5.4 sec horizontal; 2 sec torsion
Design acceleration	15 mg peak, 10-yr return
Design damping	$1\frac{1}{2}$% serviceability; $2\frac{1}{2}$% ultimate
Earthquake loading	Not applicable
Type of structure	Shear walls with outriggers at transfer levels and interior diagonals in office levels
Foundation conditions	Rock, 4-MPa (40-ton/ft^2) capacity
Footing type	Spread footings
Typical floor	
Story height	3.5 m (11 ft 6 in.) office; 2.85, 2.95, 3.05 m (9 ft 4 in., 9 ft 8 in., 10 ft) residential
Beam span, spacing	Vary
Beam depth	508 mm (20 in.) at perimeter
Slab	Flat slab
Thickness	216 mm (8.5 in.) office; 241, 267, 305 mm (9.5, 10.5, 12 in.) residential
Columns	Size and spacing vary
Material	56 MPa (8000 psi)
Core	Concrete walls of varying thickness

CitySpire, 156 West 56th Street, displaced Metropolitan Tower as the tallest concrete structure in New York City. Concrete placement reached to 244 m (800 ft) and aluminum-dome fins extended the height to 248 m (814 ft) above grade. When completed in 1987, it was the second tallest concrete structure in the world (Fig. 4.61). With a 10:1 ratio, it is the tallest, most slender structure (concrete or steel) in the world today. CitySpire has about 77,100 m^2 (830,000 ft^2) of floor space and required 33,000 m^3 (43,000 yd^3) of concrete and 4300 tonnes (4700 tons) of reinforcing bars for its 77 construction levels (including mechanical and below-grade levels).

The critical wind direction for this building is from the west, which produces maximum cross-wind action in the short (north-south) direction. Wind tunnel studies indicated possible resonance with wind forces, which could cause vortex shedding to interlock with the structure and increase wind loading as well as occupant perception of acceleration. This possibility was eliminated by adding some stiffness and mass to the structure.

Fig. 4.61 CitySpire, New York, under construction. (*Courtesy of Robert Rosenwasser Assoc.*)

The modeling of CitySpire was complex because the structure is subdivided into nine major structural subsystems with many setbacks. An in-house developed computer program, TOWER, was used 15 different times to fine-tune the project and help accommodate last-minute revisions (some implemented after the construction started). Finite-element analysis (using SAP4) was also run as a backup to the calculations made by TOWER.

The main structural system is a shear-wall–open-tube system, which traverses the center 24.4-m (80-ft)-wide octagon in each direction. Staggered rectangular concrete panels were used to form space diagonals in the lower office levels. These panels occurred on a few preselected office levels to provide continuity between the 1.7- by 2.1-m (5.5- by 7-ft) jumbo columns located on the north octagon face and the center residential-elevator core. The east and west octagon columns were similarly connected by staggered concrete panels. The available open office floor space was only somewhat reduced by these panels. Access routes in both the east-west and the north-south directions and open panoramic views (essential for Manhattan occupancy) for the office and residential levels were thus provided. The apartment levels above the twenty-sixth floor required numerous coupling beams to connect the many parts of the shear wall. Above the sixty-second floor the wings, which extend from the 24.4-m (80-ft) octagon center, were eliminated. A large open span, free of supports, was maintained between the exterior and the center elevator core to accommodate flexible duplex and penthouse layouts (Fig. 4.62).

The design problems were further complicated by the inclusion of a City Center Theatre stage extension into the structure and providing glass-cleaning access (for the octagon north and south faces) from the structure's interior. Any concrete columns at the periphery could not be too wide (to allow for hand-reached cleaning distance) and the beams not too deep (so that views are not obstructed). Some of the structural solutions to such restrictions, especially at the many setbacks requiring large column load transfers, are evident in Fig. 4.61. At each of the setbacks, transfer girders joined other belt beams to mobilize, via Vierendeel and outrigger action, additional supports and tie down loads to the shear-wall–open-tube lateral resisting system.

The concrete strength of the columns varied between 57 and 39 MPa (8300 and 5600 psi). The corresponding strength of the floor members, respecting the allowable ratio of 1.4 of the American Concrete Institute (ACI) building code, varied between 41 and 28 MPa (5950 and 4000 psi). The prefabricated cladding for this structure made special handling of the construction process necessary to compensate for the elastic and initial long-term (creep and shrinkage) shortening of the concrete supports. The concrete contractor was instructed to build in an extra 3 mm ($\frac{1}{8}$ in.) per floor to compensate for the initial construction losses. Ample soft joints were provided for future long-term creep and shrinkage losses, including racking and temperature demands.

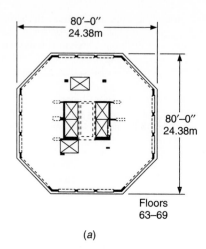

Floors
63–69

(a)

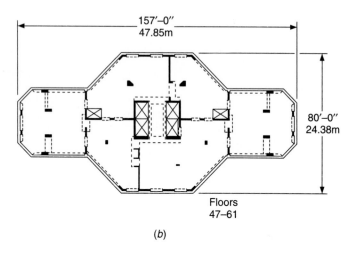

Floors
47–61

(b)

Fig. 4.62   Floor plans; CitySpire.

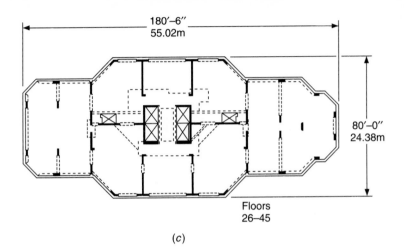

Floors
26–45

(c)

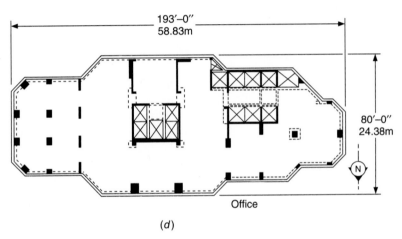

Office

(d)

Fig. 4.62   Floor plans; CitySpire. (*Continued*)

Chifley Tower
Sydney, Australia

Architect	Kohn, Pedersen, Fox with Travis Partners
Structural engineer	Flack and Kurtz Australia with Thornton-Tomasetti
Year of completion	1992
Height from street to roof	215 m (705 ft)
Number of stories	50
Number of levels below ground	4
Building use	Office with 2 retail levels
Frame material	Steel
Typical floor live load	3 kPa (60 psf)
Basic wind velocity	50 m/sec (112 mph) ultimate, 1000-yr return
Maximum lateral deflection	$H/400$, 50-yr return
Design fundamental period	5.0 sec
Design acceleration	20 mg peak, 5-yr return, with operating tuned mass damper
Design damping	2 to 2.5% serviceability; 6% ultimate
Earthquake loading	Not applicable
Type of structure	Steel perimeter frames, braced steel core with outriggers at levels 5, 29–30, 42–43
Foundation conditions	Sandstone, 5-MPa (50-ton/ft^2) capacity
Footing type	Spread footings plus rock anchors up to 18 m (60 ft) long
Typical floor	
Story height	4.075 m (13 ft 4 in.)
Beam span	10 to 15 m (33 to 49 ft)
Beam depth	530 mm (21 in.)
Beam spacing	2.5 to 3 m (8 ft 2 in. to 9 ft 10 in.)
Material	Steel, grade 350 MPa (50 ksi)
Columns	Braced steel frame
Material	Steel, grade 250 and 350 MPa (36 and 50 ksi)
Core	Braced steel frame, grade 350 MPa (50 ksi)

Chifley Tower has been designed to house financial service organizations. Wiring needs were met by raised "computer" flooring, by generous riser closets, and by the open nature of a steel-framed core. (Less accessible concrete cores are most commonly used in Australia.) Steel framing was also used to speed erection and occupancy (Fig. 4.63).

Its 90,000-m^2 (969,000-ft^2) tower rises from a 32,000-m^2 (345,000-ft^2) full-site "podium." The building has a highly articulated facade with nonparallel sides, setbacks at different levels on different elevations, and a mix of flat, gently curved, and circular

Fig. 4.63   Chifley Tower, Sydney, Australia.

faces. This design serves to define and enclose Chifley Square, reflect the street grid, maximize the prime views of harbor, park, and ocean to the north and east, break up the bulk of the tower, and enliven the Sydney skyline.

The numerous setbacks, the variety of facade geometries, and the desire for open views made a framed-tube structural solution impractical. A braced core would avoid involvement with the facade, but the tapered nature of the tower floor plans resulted in an inverted T-shape core plan (stepping back to an L at level 31) whose limited width would require unreasonably large columns to control deflection (Fig. 4.64). To control deflections more efficiently, outriggers (or heavy trusses) link the core to perimeter columns at levels 5, 29–30, and 42 (top) in the east-west direction and at levels 5 and 42 in the north-south direction. The middle east-west outriggers also serve as transfer trusses for a setback.

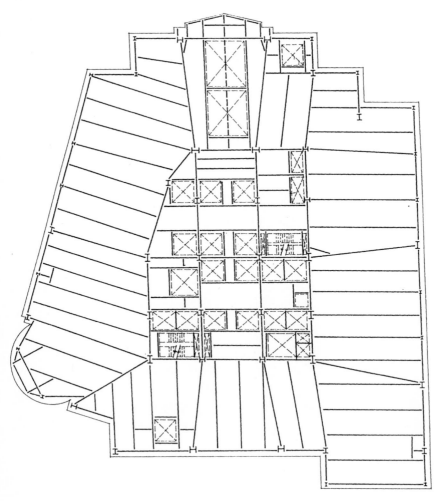

Fig. 4.64   Low-rise floor plan; Chifley Tower.

The irregular building shape, irregular core geometry, and involvement of outriggers required analyzing and designing the wind system structure by means of a complete three-dimensional computer model since no planes of symmetry exist and three-dimensional interaction was critical.

A package of analysis-and-design programs was developed for this project. An interactive deflection control routine determined "optimal" member areas to meet drift criteria by using virtual work, establishing relative efficiencies of members, resizing the most efficient members to meet deflection limits, and reanalyzing. A "final" analysis with optimal areas used precise loadings. Another analysis investigated dead loads applied to the incomplete structure under construction.

A load combination program took the member-force results of these runs and applied forces following an "overturning wind envelope" using directionality from wind tunnel tests, selected maximum and minimum wind forces for each member, and used combinations of the load cases to determine maximum design forces for each member. Wind allowable stress increases (force reductions) were included.

A member selection program used the "optimal" areas, the design forces, and a table of acceptable member sizes to select a trial member size, with an area that was near "optimal," in order to check the load capacity in accordance with the Australian steel code AS 1250-1981. The loop was then repeated with a larger trial size if necessary. Member selection marks were plotted on diagrams of the core bracing for ease of use. Member forces were also plotted in various ways to aid in the design of connections.

It is interesting to note that office dead load plus reduced live load is about 20 to 25% higher in Australia than in U.S. practice, so extrapolating U.S. tonnage figures to Australian projects could be misleading unless factored up. Australian practice also affected the construction details. Available hot-rolled member sizes are more limited than in the United States. For floor beams this meant using a heavier size than one might otherwise choose. As a result floors have a higher-than-minimum load capacity. For girders, built-up sections were common. Also, since the available plate is 100 mm (4 in.) thick or less, the largest column sections use flanges and web of doubled and tripled plates.

Chifley Tower includes a tuned mass damper (TMD) in the original construction to keep building movement below objectionable levels. Its help is not considered in the wind response for strength. The TMD mass is 400 tonnes (440 tons) of steel plate, suspended from eight 11-m (36-ft)-long cables anchored at level 46. Its period is adjusted by a tuning frame, which slides along the cables to vary their active length. Damping is provided by eight hydraulic cylinders which push fluid through a control valve and a heat exchanger in a closed circuit. Movement is permitted in any lateral direction (NSEW), but torsion is restricted by an antiyaw yoke. The TMD is anticipated to increase damping from 1 to 2.5% and to decrease 5-yr acceleration from 0.03 to 0.02 g.

One Liberty Place
Philadelphia, Pennsylvania, USA

Architect	Murphy Jahn
Structural engineer	Thornton-Tomasetti Engineers
Year of completion	1988
Height from street to roof	288 m (945 ft)
Number of stories	61
Number of levels below ground	1
Building use	Office
Frame material	Structural steel-braced core with super-diagonal outriggers
Typical floor live load	2.5 kPa (50 psf)
Basic wind velocity	31 m/sec (70 mph)
Maximum lateral deflection	$H/450$
Design fundamental period	5.5 sec
Design acceleration	15 mg peak, 10-yr return
Design damping	1 to 2%
Earthquake loading	Not applicable
Type of structure	Braced steel core linked by steel girders to exterior columns
Foundation conditions	Rock, 4-MPa (40-ton/ft^2) capacity
Footing type	Caissons
Typical floor	
Story height	3.81 m (12 ft 6 in.)
Beam span	13.4 m (44 ft)
Beam depth	530 mm (21 in.)
Beam spacing	3.05 m (10 ft)
Material	Steel, grade 350 MPa (50 ksi)
Slab	63-mm (2.5-in.) concrete over 76-mm (3-in.) metal deck
Columns	
Size at ground floor	W350 by 384 (W350 by 257) built up to 2788 kg/m (1870 lb/ft)
Spacing	6.1, 13.4, 21.3 m (20, 44, 70 ft)
Core	Linked braced frame with outriggers
Material	Steel, grade 250-MPa (36-ksi) bracing, grade 300-MPa (43-ksi) and 350-MPa (50-ksi) beams and columns

One Liberty Place at 288 m (945 ft) is located on a prime block of downtown Philadelphia (Fig. 4.65). The office floors range from 2230 m^2 (24,000 ft^2) in the lower portions to 120 m^2 (1300 ft^2) at the peak. The 61-story tower contains over 120,000 m^2 (1.3 million ft^2) of floor area.

Fig. 4.65   One Liberty Place, Philadelphia, Pennsylvania.

Structural steel framing was chosen for its flexibility and high strength—in particular, its ability to transmit large tensile and compressive forces efficiently while keeping the size of the members to a minimum. Built-up wide-flange sections were used for all outrigger diagonals and core and outrigger columns due to the large forces and required thickness of the plates. Their use also facilitated fabrication and erection.

The typical floor framing consists of composite W21 ASTM A-572 grade 50 steel beams spanning 13.4 m (44 ft) from the building core to the exterior face. As a result, the entire lease space within the tower is column-free (Fig. 4.66). The structural slab is composed of a 76-mm (3-in.) composite decking with 64-mm (2.5-in.) stone concrete topping. Floor beams were cambered to compensate for dead-load deflection under wet concrete placement.

The selected lateral load resisting system is a superdiagonal outrigger scheme composed of a 21.3- by 21.3-m (70- by 70-ft) braced core coupled with six four-story diagonal outriggers at each face of the core located at three points over the height of the building. The system works in a similar manner to the mast of a sailboat, with the braced core acting as the mast and the outrigger superdiagonals and verticals forming the spreader

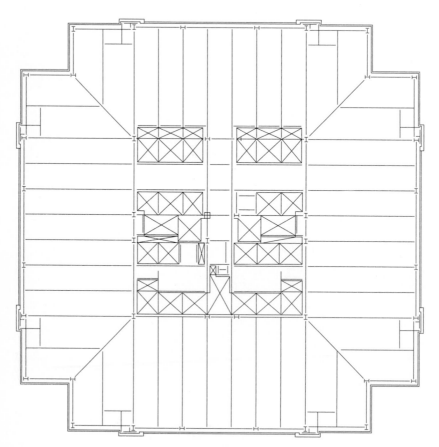

Fig. 4.66   Typical floor plan; One Liberty Place.

and shroud system. After various studies utilizing in-house optimization computer programs, three sets of eight outriggers were found to be the most efficient solution.

Although simplified models showed that they would be the most effective if spaced at equal intervals, optimization programs showed that these outriggers could further reduce wind-induced drift without adding additional steel by simply modifying their spacing over the height of the building. Ultimately the design was completed with the outside ends of the superdiagonals placed at floors 20, 37, and 51. The outrigger superdiagonals are connected at the exterior of the building to vertical outrigger columns.

To reduce uplift forces on corner core columns and the outrigger columns it was desirable to concentrate most of the building's dead load on these columns. This was accomplished by introducing exterior transfer trusses at floors 6, 21, and 37, which span between the outrigger columns within the exterior face and thus funnel dead load into the outrigger columns to compensate for uplift due to wind pressure. Uplift in the exterior outrigger columns was totally eliminated with this approach. The uplift on the corner core columns was reduced to 5800 kN (1,300,000 lb).

In developing the superdiagonal outrigger system, an intensive effort between the building's architects, interior planners, and developer was undertaken to determine that the presence of diagonal outriggers penetrating down through certain lease space at eight locations on 12 floors would not interfere with the efficient layout of the space. Interior planners made various layouts for full-floor and partial-floor tenants and concluded that the presence of the inclined superdiagonal columns would not hinder the real estate leaseability of these spaces.

Wind forces were generated using prevailing codes and also utilizing a force-balance wind tunnel test undertaken by Cermak/Peterka of Fort Collins, Colorado. It was determined that average wind pressures on the building varied between 0.25 kPa (5 psf) at the bottom to 2.9 kPa (58 psf) at the top. Both planar and three-dimensional static and dynamic analyses were performed for combinations of gravity and lateral loads. The period of the building was determined to be 5.5 sec.

The lateral load resisting system was initially designed using a purely allowable stress criterion. During the optimization effort, members were increased in size, which contributed to increasing the building's internal stiffness. As stiffness was increased, the acceptable limits of building drift ($H/450$) and acceleration (15 mg) were met. In addition, because of the vertical compatibility between outrigger columns and core columns created by the outriggers, analyses were required to determine the gravity load magnitude in the lateral load resisting system. This analysis was performed in steps to properly model the actual building erection and loading sequences.

Utilization of the optimization program trimmed an estimated 9.8 kg/m^2 (2 psf) from the wind-resisting system, a savings of some 15% by weight. More important were the savings gained by eliminating entire components such as two interior bracing lines above the twentieth floor, which greatly simplified design and construction.

17 State Street
New York, N.Y., USA

Architect	Emery Roth and Sons
Structural engineer	Desimone, Chaplin and Dobryn
Year of completion	1988
Height from street to roof	167.3 m (542 ft 2 in.)
Number of stories	44
Number of levels below ground	1
Building use	Office
Frame material	Steel
Typical floor live load	2.5 kPa (50 psf)
Basic wind velocity	47 m/sec (105 mph), 100-yr return
Maximum lateral deflection	H/500, 100-yr return
Design fundamental period	4.7, 5.0 sec
Design acceleration	20 mg peak
Design damping	1% serviceability
Earthquake loading	Not applicable
Type of structure	Bundled braced core tubes with perimeter moment frame and an outrigger hat truss
Foundation conditions	Rock
Footing type	Concrete piers and steel piles
Typical floor	
Story height	3.66 m (12 ft)
Beam span	5.5 to 12.2 m (18 to 40 ft)
Beam depth	305 to 530 mm (12 to 21 in.)
Beam spacing	3.2 m (10 ft 6 in.) max
Slab	63-mm (2.5-in.) normal-weight concrete on 76-mm (3-in.) metal deck
Columns	Built-up W350 (W14) core, W610 (W24) perimeter
Spacing	8.53 m (28 ft) core, 5.69 m (18 ft 8 in.) perimeter
Material	Steel, grade 250 MPa (36 ksi)
Core	Braced tubes, grade 250-MPa (36-ksi) concrete encased through lowest two levels

17 State Street is a 44-story office tower located across from Battery Park at the tip of Manhattan (Fig. 4.67). To maximize the unobstructed views of the Statue of Liberty and the New York harbor, the architects chose a quarter-circle floor plan of 1160 m^2 (12,500 ft^2) (Fig. 4.68). Although the perimeter of the plan is symmetric, the core of the tower is offset to optimize the arrangement of rental floor space. The first level is 10 m (33 ft) above grade, and typical floors are 3.66 m (12 ft) high.

Fig. 4.67   17 State Street, New York.

Wind tunnel testing predicted that the wind coming off the harbor would produce loads 40% higher than those required by the New York City building code.

The structural system consists of bundled braced core tubes coupled to perimeter moment frames by means of an outrigger hat truss. The three core tubes are braced with X, diagonal, and inverted V members, as dictated by core functional requirements. Core

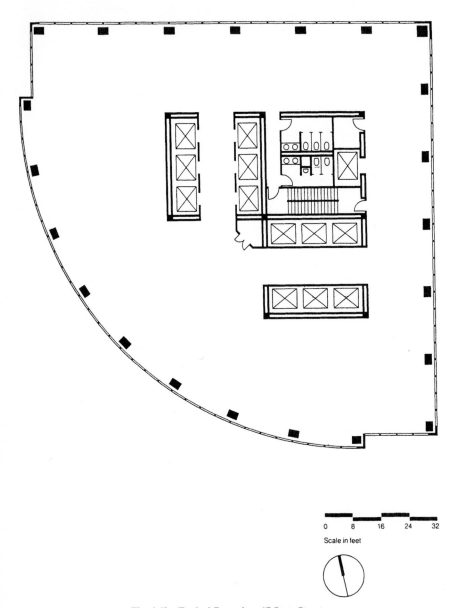

0   8   16   24   32

Scale in feet

Fig. 4.68   Typical floor plan; 17 State Street.

columns consist of W350 (W14) series rolled shapes in the upper portion of the building and built-up members below. Perimeter moment frames have W610 (W24) series columns, rolled and built up, spaced at 5.7 m (18 ft 8 in.). The perimeter frames do not form a tube, as architectural notches at the corners of the quarter-circle prevent effective economical transfer of vertical shear forces around the corners. The hat truss is a three-dimensional outrigger two stories high, with diagonals sloping downward from the core to the perimeter.

At the first level, which rises 10 m (33 ft) above the sidewalk, the perimeter columns and spandrel beams are encased in concrete to provide additional stiffness for the tall story. Below the ground-floor level, the cores are also encased to add stiffness. Footings consist of concrete piers to 6-MPa (60-ton/ft^2) bedrock and end-bearing steel piles. Eight columns are anchored for uplift with posttensioned threadbar rock anchors.

Figueroa at Wilshire
Los Angeles, California, USA

Architect	Albert C. Martin
Structural engineer	CBM Engineers, Inc.
Year of completion	1990
Height from street to roof	218.5 m (717 ft)
Number of stories	53
Number of levels below ground	4
Building use	Office
Frame material	All steel
Typical floor live load	2.5 kPa (50 psf)
Basic wind velocity	31 m/sec (70 mph)
Maximum lateral deflection	380 mm (15 in.), 100-yr return
Design fundamental period	6.5 sec
Design acceleration	17 mg peak, 10-yr return
Design damping	1% serviceability; 7% ultimate
Earthquake loading	Magnitude 8.3 from San Andreas fault
Type of structure	Braced core "spine" with outrigger ductile frame
Foundation conditions	Shale, 750-kPa (15,000-psf) capacity
Footing type	Spread footings
Typical floor	
Story height	3.96 m (13 ft)
Beam span	18.3 to 10.7 m (60 to 35 ft)
Beam depth	914 to 406 mm (36 to 16 in.)
Beam spacing	3.05 m (10 ft)
Slab	133-mm (5.25-in.) lightweight concrete on 50-mm (2-in.) metal deck
Columns	
Size at ground floor	1067 by 1067 mm (42 by 42 in.), cruciform shape at 18.3-m (60-ft) centers
Material	Steel, grade A572, 350 MPa (50 ksi)
Core	Braced steel, grade A572

This 218.5-mm (717-ft)-tall 53-story office tower is located in downtown Los Angeles (Fig. 4.69). The floor plan of the tower is 45.7 m (150 ft) square, exhibiting notches and multiple step backs as it rises above the plaza (Fig. 4.70). The square tower plan offers internal space appropriate to banking and law firms. The granite-clad building has a three-story-tall stepped green-colored glass crown, which is lit from within at night and makes a distinct mark on the Los Angeles skyline. Two six-story atriums, both rectangular in plan, which rise like glass and steel staircases, are attached to two of the building's corners at 45° angles. The plaza of the tower at the corner of Figueroa at Wilshire is articulated by fountains and a 12-m (40-ft)-high sculpture.

Fig. 4.69   Figueroa at Wilshire, Los Angeles, California. (*Courtesy of CBM Engineers, Inc.*)

As opposed to conventional perimeter ductile tubular frames, the concept of a spine structure is used for this tower. The spine, the uninterrupted portion of this tower, consists of a 17.4- by 20.4-mm (57- by 67-ft) concentrically braced core linked to perimeter columns by a ductile frame of outrigger beams. The spine in this case has three components (Fig. 4.71):

1. A rectangular concentrically braced core anchored at its extremities by steel columns of a maximum size of 1067 mm (42 in.) square at their base. The interior core bracing and beams are proportioned in such a way that, in case of an inadvertent failure of the diagonals, the vertical load-carrying ability of the floor is not affected.

2. Outrigger beams linking the internally braced core to the perimeter columns. These beams not only carry the floor loads, but along with the perimeter columns perform the function of ductile moment resisting frames for the entire structure. The beams are laterally braced to prevent lateral torsional buckling and are con-

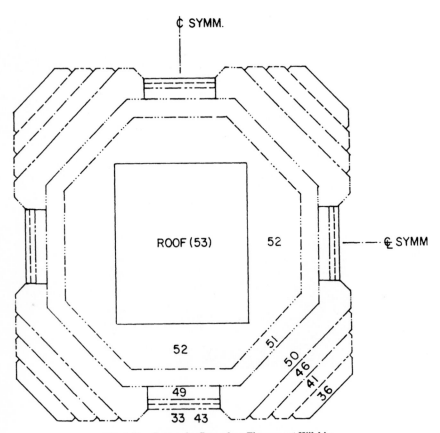

Fig. 4.70   Composite floor plan; Figueroa at Wilshire.

nected to floor diaphragms by shear studs to transmit horizontal shear forces to the frame. Notches at the midspan of those beams, which provide for the passage of mechanical ducts, are stiffened to prevent the formation of a three-hinge mechanism when the ends of beams yield during a major seismic event.

3. The 914- by 762-mm (36- by 30-in.) steel perimeter columns which, because of their importance in the overall stability of the frame, are checked for the loads created by the plastification of all outrigger beams.

Because of the closeness of lateral periods of vibrations with torsional vibration periods, the structure was checked for the phenomenon of modal coupling.

The spine structure not only provided column-free uninterrupted lease spaces, but also was structurally very efficient. Designed to remain essentially elastic for the maximum credible earthquake, the structure uses 110 kg/m^2 (22.5 psf) of structural steel, as opposed to a conventional ductile frame, which would have required 132 kg/m^2 (27 psf).

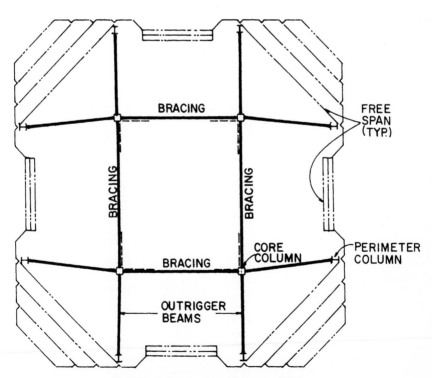

Fig. 4.71   Spine structure; Figueroa at Wilshire.

Four Allen Center
Houston, Texas, USA

Architect	Lloyd Jones Brewer Associates
Structural engineer	Ellisor and Tanner, Inc.
Year of completion	1984
Height from street to roof	210.5 m (690 ft 8 in.)
Number of stories	50
Number of levels below ground	2
Building use	Office
Frame material	Steel
Typical floor live load	2.5 kPa (50 psf)
Basic wind velocity	41 m/sec (92 mph)
Maximum lateral deflection	$H/400$, 50-yr return
Design fundamental period	4.03 sec
Earthquake loading	Not applicable
Type of structure	Braced steel core with outriggers to steel perimeter framed tube
Foundation conditions	Deep stiff clay
Footing type	Continuous mat
Typical floor	
Story height	3.96 m (13 ft)
Beam span	12.2 m (40 ft)
Beam spacing	4.57 m (15 ft)
Beam depth	610 and 915 mm (24 and 36 in.)
Material	Steel, grade 250 MPa (36 ksi)
Slab	82-mm (3.25-in.) lightweight concrete on 76-mm (3-in.) steel deck
Columns	
Size at ground level	915 by 280 mm (36 by 11 in.)
Spacing	4.57 m (15 ft)
Material	Steel, grade 250 MPa (36 ksi)
Core	Braced steel frame, grade 250 MPa (36 ksi)

The Four Allen Center building rises 50 stories above grade and extends two stories below (Fig. 4.72). The elongated plan, combined with the slenderness of the tower, yields an illusion of exceptional height when viewed from street level. The 133,800 m^2 (1.44 million ft^2) office building is connected to parking and retail facilities by an air-conditioned pedestrian tunnel and an overhead pedestrian bridge. Figure 4.73 shows the typical floor framing plan, and Fig. 4.74 illustrates the building section of a typical floor.

The geometry of the slender airfoil shape is susceptible to dynamic oscillation in hurricane-speed winds, thereby establishing a complex and challenging series of structural frame and foundation problems. Wind tunnel tests of an aeroelastic model of the building were recommended and coordinated by the structural engineers. The testing resulted

Fig. 4.72   Four Allen Center, Houston, Texas.

in developing a lateral wind-resisting system to control predicted dynamic oscillation of the building.

A four-celled tube structure was developed which includes a perimeter framed tube and three vertical trusses transverse to the elevator core, linked to the perimeter tube by tree-beam elements. The unique wind-bracing system was subjected to a full-scale test during hurricane Alicia in August 1983, and it performed exceptionally well.

A refinement of the traditional solider pile was developed to retain the 11.3-m (37-ft)-deep foundation excavation. The improved shape reduced the number of piles and

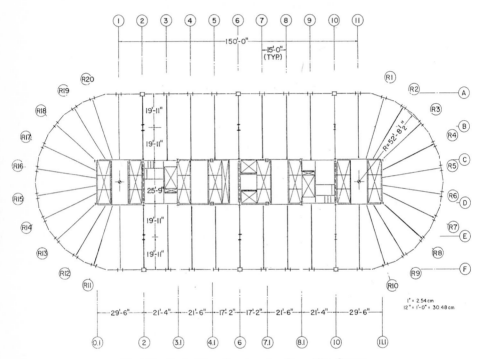

Fig. 4.73   Typical floor framing plan; Four Allen Center.

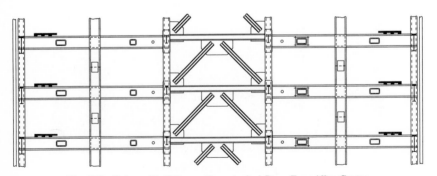

Fig. 4.74   Enlarged building section—typical floor; Four Allen Center.

tiebacks normally required, thus enhancing economy and shortening the schedule for the basement and foundation construction.

The structural development, system analysis, and design were facilitated by developing a comprehensive series of computer analyses and design programs. The automated analysis and design processing of all elements in the wind-resistant system of the building structure resulted in significant savings in material costs, and enabled the engineers to complete the design and drawings in a short 4-month schedule.

Advanced methods were also employed to assure quality control during construction. In particular, the project set new standards of assurance regarding the tightness of high-strength bolts. Ultrasonic extensometers were used to measure bolt tightness accurately for the first time on a commercial project.

The 45.7- by 91.4- by 2.6-m (150- by 300- by 8.5-ft) mat foundation containing 11,127 m^3 (13,308 yd^3) of concrete was poured in just over 19 hours. This was made possible by using a system of belt conveyors supplemented by concrete pumps.

The structural steel was erected by fabricating the exterior tree columns, the vertical core trusses, and the tree beams in modules to reduce the number of pieces to handle and field connections to complete. The all-steel structure was erected at a rate of one complete floor every $2\frac{1}{2}$ days. The project was complete 6 months ahead of the planned fast-track design and construction completion date, with the first tenant moved in just 15 months after construction of the foundation began.

The project received the following awards:

- "One of the Ten Outstanding Engineering Achievements in the United States of America," National Society of Professional Engineers, 1983
- "Grand Award Winner for High Professional Execution of Engineering Design," American Consulting Engineers' Council, 1984
- "Eminent Conceptor Award for the Most Outstanding Engineering Project," Consulting Engineers' Council of Texas, 1984

Trump Tower
New York, New York, USA

Architect	Swanke Hayden Connell
Structural engineer	Office of Irwin G. Cantor
Year of completion	1982
Height from street to roof	202 m (664 ft)
Number of stories	58
Number of levels below ground	3
Building use	Retail, offices, residential
Frame material	Concrete
Typical floor live load	5 kPa (100 psf) retail; 2.5 kPa (50 psf) offices; 2 kPa (40 psf) residential
Basic wind velocity	Unavailable; force = 1.0, 1.25, 1.5 kPa (20, 25, 30 psf)
Maximum lateral deflection	$H/600$, 100-yr return
Design fundamental period	5.2 sec
Design acceleration	16.5 mg peak, 10-yr return
Design damping	1.5%
Earthquake loading	Not applicable
Type of structure	Concrete shear core linked by concrete outrigger walls to concrete perimeter frames
Foundation conditions	Manhattan mica schist
Footing type	Spread footings
Typical floor	
Story height	4.8, 3.66, 2.9 m (16, 12, 9.5 ft)
Slab	400-mm (16-in.) waffle slab; 190-mm (7.5-in.) flat slab
Columns	
Size at ground floor	813 by 813 mm (32 by 32 in.)
Spacing	12.2 to 7.3 m (40 to 24 ft)
Material	Concrete, 49 MPa (7000 psi)
Core	Shear walls, 457 mm (18 in.) thick at ground floor in 49-MPa (7000-psi) concrete

Trump Tower is a multiuse building occupying a prime site on 5th Avenue in New York City. Through the purchase of the air rights for adjacent sites and from bonuses awarded for the provision of public amenities, a plot ratio (building floor area to site area) of 21 was achieved, making this a very slender building.

A perimeter tube lateral load resisting system was unacceptable due to the impact of closely spaced columns on the views from the condominiums and on the shop fronts at street level. Also, structural steel was rejected due to the lead time required for supply to the site. The adopted all-concrete solution utilized concrete shear walls for lateral load resistance and deep concrete transfer girders to change the structural column grid (Fig. 4.75).

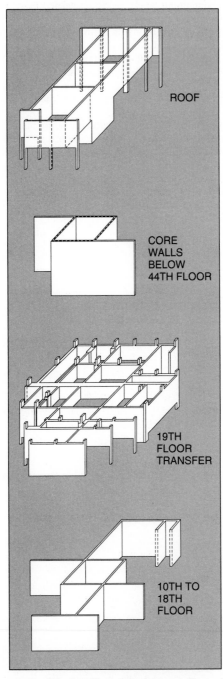

ROOF

CORE
WALLS
BELOW
44TH FLOOR

19TH
FLOOR
TRANSFER

10TH TO
18TH
FLOOR

Fig. 4.75   Column and wall load and lateral displacement; Trump Tower, New York.

Through the 38 condominium levels, loads are carried by 52 concrete columns and concrete walls around the service core. At roof level, twin outrigger beams 6 m (20 ft) high and 450 mm (18 in.) thick link the core with perimeter columns on two opposite sides to reduce lateral displacement. Extended core walls do this job in the other direction.

Below the twentieth floor a system of transfer girders 7.3 m (24 ft) high and 450 to 600 mm (18 to 24 in.) thick allows for the 52 columns to reduce to only 8 columns through the 13 office levels. The transfer girders also act as outrigger beams to further control lateral displacement. The girders are pierced by many openings for doors, pipes, and ducts.

Another transfer system comprising an inclined-column A frame was introduced between the eleventh and seventh floors to allow another two columns to be removed in order to open up the atrium, which rises seven levels through the retail floors at the base of the building.

The 1087-m^2 (11,700-ft^2) residential floors were poured on a 2-day cycle. The 56- and 49-MPa (8000- and 7000-psi) concrete for the columns contained a superplasticizer to increase workability for placing around dense reinforcement. Tight management of concrete deliveries was required to ensure that high-strength concrete was available at the right time for placement in slabs over and around columns.

Waterfront Place
Brisbane, Australia

Architect	Cameron Chisholm and Nicol (Qld.) Pty. Ltd.
Structural engineer	Bornhorst and Ward Pty. Ltd.
Year of completion	1990
Height from street to roof	158 m (518 ft)
Number of stories	40
Number of levels below ground	2
Building use	Office
Frame material	Concrete
Typical floor live load	3 kPa (60 psf)
Basic wind velocity	49 m/sec (110 mph)
Maximum lateral deflection	185 mm (7.25 in.), 50-yr return
Design fundamental period	5 sec
Design acceleration	2.3 mg (standard deviation), 5-yr return
Design damping	1% serviceability; 5% ultimate
Earthquake loading	Not applicable
Type of structure	Shear core with outriggers to perimeter columns
Foundation conditions	17 m (56 ft) of soft clay over rock
Footing type	25-m (82-ft)-long 1.5-m (5-ft)-diameter bored piles socketed and belled in rock
Typical floor	
Story height	3.6 m (12 ft)
Beam span	11.5 m (38 ft)
Beam depth	420 mm (16.5 in.)
Beam spacing	6.8 m (22 ft 4 in.)
Material	Posttensioned concrete
Slab	130-mm (5-in.) reinforced concrete
Columns	
Size at ground floor	1350 mm (53 in.) in diameter
Spacing	6.8 m (22 ft 4 in.)
Material	Concrete, 60 to 35 MPa (8600 to 5000 psi)
Core	Concrete shear walls
Thickness at ground floor	600 mm (24 in.) max
Material	Concrete, 45 to 35 MPa (6400 to 5000 psi)

Waterfront Place is a 42-level reinforced concrete framed office tower, located at the river edge of Brisbane's central business district on a 15,000-m^2 (160,000-ft^2) site (Fig. 4.76). A street-level plaza provides access to the river edge for the public, whereas below and above river level there is parking for 500 cars. River edge boardwalks connect

Fig. 4.76   Waterfront Place, Brisbane, Australia.

neighboring developments with the plaza, and mooring is provided on the waterfront for pleasurecraft, tour boats, and ferries.

The 42-level tower provides 36 office floors, three plant-room floors, a ground-floor foyer, and two basements. The configuration of a typical floor provides 12 m (40 ft) of column-free space between core and glass line, with four cantilevered bay windows on both the east and west facades, effectively contributing 10 "corner" windows on each floor (Fig. 4.77).

High-rise buildings taller than approximately 35 stories may not be structurally economical if the core alone is used to resist wind loads. This is particularly the case for a building rectangular in plan loaded about its weak axis. Such was the case with Waterfront Place, which has 40 levels above plaza level.

Wind tunnel model testing was undertaken, and the results indicated that it would be impractical to use the core to fully withstand wind forces. Wall thicknesses and reinforcement quantities would be excessive, as would be the sway of the building in the east-west direction.

Instead the design concept was changed to that of a core-perimeter interaction structural system where the core "tube" is connected to the exterior columns at specific locations, in this case at the plant room at levels 26 and 27 (Fig. 4.78). At these levels, four stiff "wind beams" cantilevering from the core are connected to perimeter transfer beams between three columns on each face of the building. This induces participation of the axial capacity of the exterior columns in resisting wind-induced loading (Fig. 4.79).

The core is used to resist all horizontal shear, but vertical shear resistance is transferred from the core to the exterior columns, thereby utilizing the total overturning capacity of the structure.

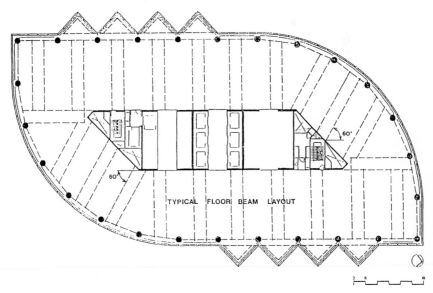

TYPICAL FLOOR BEAM LAYOUT

Fig. 4.77   Typical midrise floor plan; Waterfront Place.

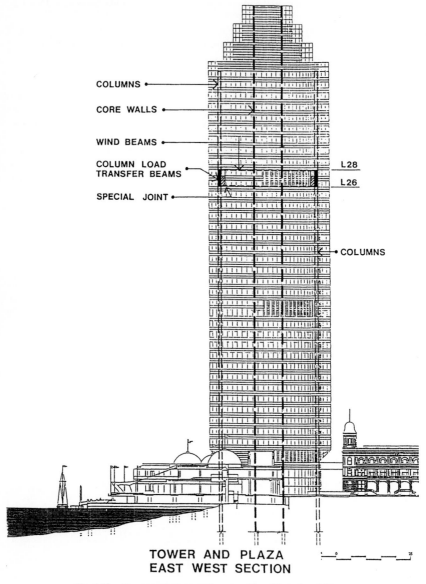

COLUMNS •

CORE WALLS •

WIND BEAMS •

COLUMN LOAD
TRANSFER BEAMS •   L28

                                   L26

SPECIAL JOINT •

                                   • COLUMNS

**TOWER AND PLAZA
EAST WEST SECTION**

Fig. 4.78   Tower and plaza east-west section; Waterfront Place.

Research indicated that the most effective location for the wind beams was at the top levels of the tower. However, this was impractical due to the stepped profile of the topmost three plant levels (levels 37, 38, and 39) and the marketing potential of tenancy levels 29 to 36. As a consequence the wind beams were placed at levels 26 and 27, two floors containing mechanical rooms and office space.

This location heightened the possibility of differential axial shortening between the reinforced concrete core and the columns. A steel joint was developed to link the outrigger beams with the transfer beams at the columns to allow controlled slippage as the differential movement occurred.

The use of the cantilevering wind beam system introduced some architectural and structural engineering design challenges. In order to resist the 820-tonne (180,000-lb) load applied to the end of each wind beam, the beams had to be two stories high and 900 mm (36 in.) thick and preferably without any penetrations. To have no penetrations would have meant the loss of office space; therefore large openings were made in these beams. This precluded the use of conventional beam design theory for these beams. Consequently the beams were designed using "strut and tie" theory. Concrete of 55-MPa (7800-psi) strength and ties consisting of 45 36-mm (1.4-in.)-diameter bars were required to transmit the working load of 820 tonnes (180,000 lb) per beam.

The floor slabs at levels 26 and 28, which are locally 420 mm (16.5 in.) thick, participate in the wind-beam action by working as flanges for the wind beams. The force paths in the wind beams and the floor slabs are shown in Figs. 4.79, 4.80, and 4.81.

Differential vertical shrinkage between core and perimeter columns at level 26 subsequent to construction of the entire building was calculated. Construction history, material properties, and in-service loads were used in this calculation.

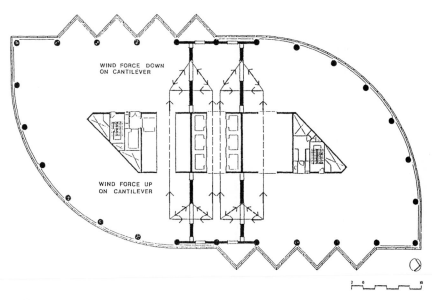

Fig. 4.79   Level 26 floor plan—forces transmitted through floor "flange"; Waterfront Place.

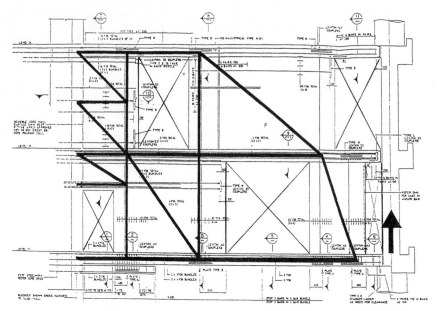

Fig. 4.80   Strut/tie truss—force up; Waterfront Place.

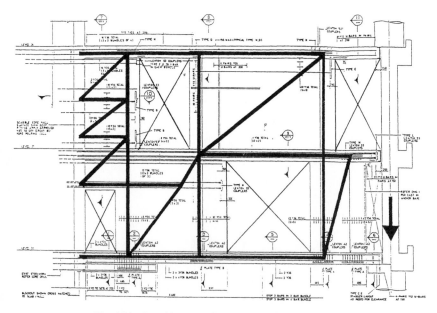

Fig. 4.81   Strut/tie truss—force down; Waterfront Place.

The wind beams are extremely stiff. Design load deflection was calculated to be only 2 mm (0.08 in.). Unless some means of allowing movement between wind beam and columns was found, the wind beams would have attempted to support the 15 stories above level 26 and several stories below. This it could not do, and structural failure would have resulted. A sliding friction joint between wind beams and the column transfer beams was developed. This is shown in Fig. 4.82. The joint is in effect a multiple clutch with the slip load determined by the clamping force provided by the through bolts.

Tests were carried out at the Queensland University of Technology to determine the co-efficient of friction between the brake-pad material and the stainless-steel plates. Size, clamping force, and loading rate effects were investigated. Typical load-slip graphs are shown in Fig. 4.83. Each joint is fitted with four strain gauges to monitor stresses in the plates and hence the load being transferred through the "clutch." This allows the clamping force to be adjusted to slip at the required design load. When the clamping force is finally adjusted, it will not require any further adjustment in its life. A typical plot of stress versus time for one of the joints is shown in Fig. 4.84.

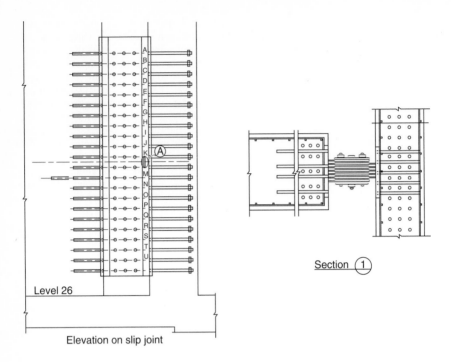

Elevation on slip joint

Section ①

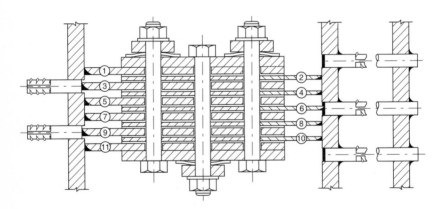

Plan detail of slip joint

Fig. 4.82   Slip joint; Waterfront Place.

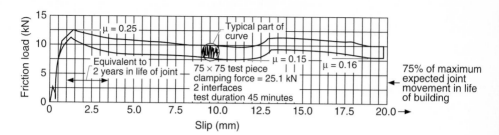

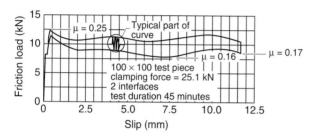

Fig. 4.83   Friction tests; Waterfront Place.

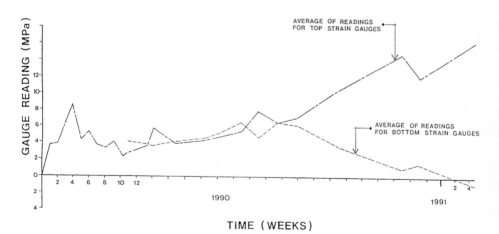

Fig. 4.84   Typical strain-gauge readings on wind-beam joint; Waterfront Place.

Two Prudential Plaza
Chicago, Illinois, USA

Architect	Loebe Schlossman and Hackl
Structural engineer	CBM Engineers, Inc.
Year of completion	1990
Height from street to roof	278 m (912 ft)
Number of stories	64
Number of levels below ground	5
Building use	Office
Frame material	Concrete to level 59, steel above
Typical floor live load	4 kPa (80 psf)
Basic wind velocity	31 m/sec (70 mph)
Maximum lateral deflection	488 and 419 mm (19.2 and 16.5 in.), 50-yr return
Design fundamental period	7.2, 5.8, 4.4 sec
Design acceleration	19 mg peak, 10-yr return
Design damping	2% serviceability
Earthquake loading	Not applicable
Type of structure	Shear core with outrigger beams and perimeter frame
Foundation conditions	14-m (45-ft) fill over 11-m (35-ft) hard-pan over rock
Footing type	15-m (50-ft) hardpan caissons and 24-m (80-ft) rock caissons
Typical floor	
Story height	3.96 m (13 ft)
Beam span	12.2 m (40 ft)
Beam depth	610 mm (24 in.)
Beam spacing	6.1 m (20 ft)
Slab	One-way 150-mm (6-in.) slabs, typically 28-MPa (4000-psi) concrete
Columns	
Size at ground floor	890 by 1140 mm (35 by 45 in.) at 6.1-m (20-ft) centers
Material	Concrete, 84 MPa (12,000 psi)
Core	Shear walls 840, 610, 460 mm (33, 24, 18 in.) thick at ground floor
Material	Concrete, 84 MPa (12,000 psi)

Two Prudential Plaza, a 64-story office building, is located in downtown Chicago, Illinois (Fig. 4.85). At the time of completion it was the second tallest concrete building in the world. The building has a gross area of about 130,000 m^2 (1.4 million ft^2). It has five levels of basement, which are primarily used as a parking garage for 325 cars. The low-

Fig. 4.85   Two Prudential Plaza, Chicago, Illinois. (*Courtesy of CBM Engineers, Inc.*)

est basement is located at elevation −6.477 m (−21 ft 3 in.) CCD (Chicago City datum). The lobby of the building is located at elevation + 10.668 m (+35 ft 0 in.) CCD. Levels 4, 5, 38, and 39 are used for mechanical equipment, and level 59 for storage of window-washing equipment. Level 58 is the last office floor. Levels above 59 are mechanical floors.

The building is rectangular at the lower levels, 37.4 by 40.4 m (122 ft 6 in. by 132 ft 8 in.), but becomes square at the fifty-ninth floor due to a series of setbacks on the north and south faces. Above the fifty-ninth floor, the building starts tapering to form a "cone head," which is topped by a 25-m (82-ft) architectural spire. The top elevation of the spire is 304.8 m (1000 ft) CCD (Fig. 4.86).

The lateral stiffness in each direction is mainly provided by the four shear walls located in the core of the building. Their depth is 13.8 m (45 ft 4 in.). The flanges are 838 mm (33 in.) thick and the webs are 610 and 380 mm (24 and 15 in.) thick for the interior and exterior walls, respectively. The south shear wall drops off at level 27 whereas the north wall drops off at level 40. The middle walls continue all the way to floor 59. The flanges of walls are connected together in the north-south direction by 686-mm (27-in.)-deep link beams.

The columns at the east and west faces are spaced at 6.1-m (20-ft) centers, whereas on the north and south faces they are spaced at 9.15 m (30 ft). The typical exterior column size varies from 890 by 1140 mm (35 by 45 in.) at the lower floors to 600 by 600 mm (24 by 24 in.) at the top floors. A maximum concrete strength of 84 MPa (12,000 psi) was used for columns and shear walls at the lower floors. The concrete strength was reduced to 42 MPa (6000 psi) at the upper floors.

The floor beams have a clear span of approximately 12 m (40 ft) from the perimeter columns to the shear wall core. Typical floor beam size is 965 mm (38 in.) by 610 mm (24 in.) deep. Floor framing consists of a 150-mm (6-in.)-thick normal-weight concrete slab with a clear span of 5.13 m (16 ft 10 in.) between the floor beams, spaced at 6.1 m (20 ft) centers. In addition to carrying the gravity load, the floor beams carry some of the wind shear from the shear walls to the outside columns. At the fortieth and the fifty-nineth floors the core is tied to the outside columns at two locations with the help of outrigger walls to control the wind drift and reduce the overturning moment in the core shear walls. The beams are 5.03 m (16 ft 6 in.) deep (in other words, a full story high) between floors 39 and 40 and 1.68 m (5 ft 5 in.) deep at floor 59.

The foundation system consists of straight shaft caissons up to 3 m (10 ft) in diameter. These caissons rest on the bedrock, which is about 30 m (100 ft) below the existing ground level. The allowable bearing capacity of this rock is 19 MPa (200 ton/ft²). To fully utilize this capacity, 56-MPa (8000-psi) concrete was used in caissons. In the parking garage adjacent to the main tower, belled caissons were used. These caissons extend to hardpan about 21 m (70 ft) below existing grade. The allowable bearing capacity for this hardpan is about 3.4 MPa (36 ton/ft²).

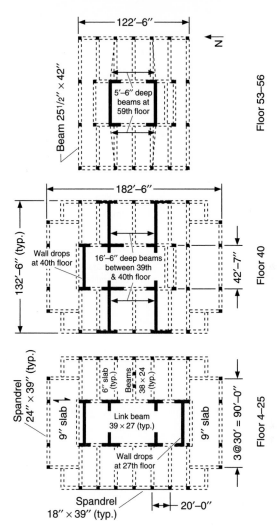

Fig. 4.86   Typical floor plans; Two Prudential Plaza.

1999 Broadway
Denver, Colorado, USA

Architect	C.W. Fentress and Associates P.C.
Structural engineer	Severud Associates
Year of completion	1985
Height from street to roof	198 m (650 ft)
Number of stories	43
Number of levels below ground	3
Building use	Office
Frame material	Concrete core, steel frame
Typical floor live load	2.5 kPa (50 psf)
Basic wind velocity	36 m/sec (80 mph)
Maximum lateral deflection	$H/400$
Earthquake loading	USA zone 1
Type of structure	Concrete core with outriggers to perimeter steel frame
Footing type	Caissons
Typical floor	
Story height	3.81 m (12 ft 6 in.)
Beam span	9.14 m (30 ft)
Beam depth	406 mm (16 in.)
Beam spacing	3.43 m (11 ft 3 in.)
Slab	83-mm (3.25-in.) lightweight concrete on 50-mm (2-in.) metal deck
Columns	
Size at ground level	W350 by 1088 kg/m (W14 by 730)
Spacing	4.57 m (15 ft)
Core	Shear walls 610 mm (24 in.) thick at ground floor
Material	Concrete, 42 to 28 MPa (6000 to 4000 psi)

1999 Broadway is an unusual 43-story office building built on a triangular site. The presence of an historic church on part of the site resulted in the plan of the office building having the shape of an arrowhead which wraps around the church, creating from it a piece of sculpture on the plaza (Fig. 4.87).

The facade comprises alternating bands of limestone and green reflective glass and a concave curtain wall having seven angled facets around and above the church. The building has been raised 15 m (50 ft) above ground on 22 limestone-clad columns to create views of the church from within.

The structure consists of a reinforced concrete service core, steel perimeter columns, and steel floor beams and girders composite with the slab. At levels 3 to 5 and 29 to 31, two-story-high outrigger trusses between core and perimeter columns reduce the lateral deflection. Girders are connected to plates field-welded to cast-in plates in the slip-formed core. The perimeter frames act with the core to resist lateral loads and effects due to the eccentric form of the building.

Footings comprise cast-in-place caissons founded in claystone and sandstone some 15 m (50 ft) below grade. A single caisson supports each column, and caissons at a minimum spacing of three caisson diameters are distributed around the core. The design end bearing pressure was 3350 kPa (70,000 psf), and skin friction in the rock was 335 kPa (7000 psf).

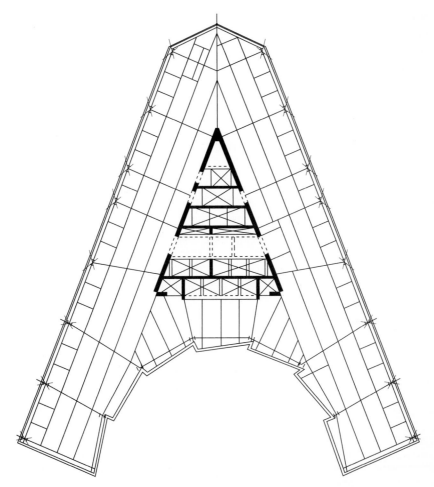

Fig. 4.87   Typical floor plan; 1999 Broadway, Denver, Colorado.

Citibank Plaza
Hong Kong

Architect	Rocco Design Partners
Structural engineer	Ove Arup and Partners
Year of completion	1992
Height from street to roof	220 m (722 ft)
Number of stories	41
Number of levels below ground	4
Building use	Office
Frame material	Reinforced and posttensioned concrete
Typical floor live load	5 kPa (104 psf)
Basic wind velocity	64 m/sec (144 mph), 50-yr return, 3-sec gust
Maximum lateral deflection	370 mm (14.5 in.), for 50-yr return period wind
Earthquake loading	Not applicable
Type of structure	Concrete core with outriggers
Foundation conditions	Decomposed granite over granite bedrock
Footing type	Hand-dug caissons to rock
Typical floor	
Story height	3.9 m (12.8 ft)
Beam span	9.4 m (31 ft)
Beam depth	500-mm (20-in.)-deep ribbed slab
Beam spacing	Reinforced concrete ribbed slab
Columns	
Size at ground floor	3000 by 1900 mm (120 by 75 in.) max
Spacing	9.4 m (31 ft)
Material	Concrete with 40-MPa (5800-psi) cube strength
Core	Shear walls 1.0 and 1.2 m (3.3 and 4 ft) thick at base
Material	Concrete with 40 MPa (5800-psi) cube strength

The four-level basement of Citibank Plaza (Fig. 4.88) was formed using top-down construction techniques. Stability was achieved with the internal cores acting in combination with the perimeter columns, using outriggers at two levels (Fig. 4.89). Part of the building is seated above a major entryway to a neighboring development. To achieve this, the perimeter columns rake outward along one face of the building over a one-story height (Fig. 4.90). The resulting lateral forces were resisted by a prestressed beam system tied back to the internal cores, prestressing being applied in stages as construction progressed.

Fig. 4.88   Citibank Plaza, Hong Kong. (*Courtesy of Ove Arup and Partners.*)

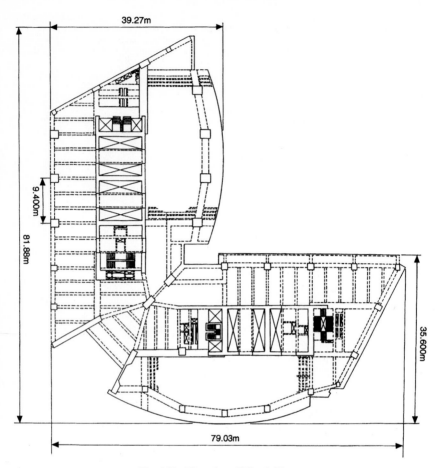

Fig. 4.89   Floorplan; Citibank Plaza.

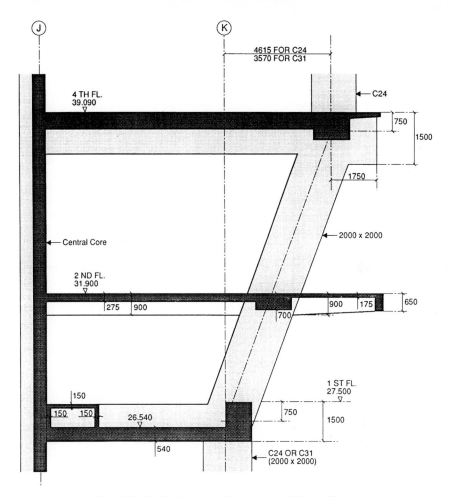

Fig. 4.90   Section through raking columns; Citibank Plaza.

4.4   TUBULAR SYSTEMS

1   Historical Perspective

The development of the initial generation of tubular systems for tall buildings can be traced to the concurrent evolution of reinforced concrete construction following World War II. Prior to the early 1960s, reinforced concrete was utilized primarily for low-rise construction of only a few stories in height. These buildings were characterized by planar Vierendeel beam and column arrangements with wide spacings between members. The basic inefficiency of the frame system for reinforced concrete buildings of more than about 15 stories resulted in member proportions of prohibitive size and structural material cost premiums, and thus such systems were economically inviable. Concrete shear wall systems arranged within the building interior could be utilized, but they were often of insufficient size for stiffness and resistance against overturning. This led to the development of structural systems with a higher degree of efficiency toward lateral load resistance for taller buildings. The notion of a fully three-dimensional structural system utilizing the entire building inertia to resist lateral loads began to emerge at this time. The main proponent of this design trend was Fazlur Khan, who systematically pursued a logical evolution of tall building structural systems. The pervasive international-style approach to architecture at the time included larger open spaces with longer spans, a well-organized core, and a clearly perceptible interior-exterior column grid. Within this architectural and economic climate, the framed tube system in reinforced concrete can be seen as both a natural and an innovative development in tall building systems.

2   The Framed Tube

The organization of the framed tube system is generally one of closely spaced exterior columns and deep spandrel beams rigidly connected together, with the entire assemblage continuous along each facade and around the building corners. The system is a logical extension of the moment resisting frame, whereby the beam and column stiffnesses are increased dramatically by reducing the clear span dimensions and increasing the member depths. The monolithic nature of reinforced concrete construction is ideally suited for such a system, involving fully continuous interconnections of the frame members. Depending on the height and dimensions of the building, exterior column spacings should be on the order of 1.5 to 4.5 m (4.9 to 14.8 ft) on center maximum. Spandrel beam depths for normal office or residential occupancy applications are typically 600 to 1200 mm (24 to 47 in.). The resulting arrangement approximates a tube cantilevered from the ground with openings punched through the tube walls. The closely spaced and deep exterior members often have a secondary benefit with respect to exterior cladding system costs. Exterior columns may eliminate the need for intermediate vertical mullion elements of the curtain wall partially or totally. A structuralist expression for the exterior envelope may be fully realized by exposing the exterior tubular members, thus defining the architectural fenestration. The window wall system is then infilled between the column and spandrel beams, with a resulting reduction in cladding cost. An early example of such a tubular building in reinforced concrete is shown in Fig. 4.91.

The behavior of framed tubes under lateral load is indicated in Fig. 4.92, which shows the distribution of axial forces in the exterior columns. The more the distribution is similar to that of a fully rigid box cantilevered at the base, the more efficient the system will be. For the case of a solid-wall tube, the distribution of axial forces would be expected to be uniform over the windward and leeward walls and linear over the sidewalls. As the tubular walls are punched, creating the beam-column frame, shear frame deformations

Fig. 4.91   Brunswick Building, Chicago, Illinois.

are introduced due to shear and flexure in the tubular members as well as rotations of the member joints. This reduces the effective stiffness of the system as a cantilever. The extent to which the actual axial load distribution in the tube columns departs from the ideal is termed the "shear lag effect." In behavioral terms, the forces in the columns toward the middle of the flange frames lag behind those nearer the corner and are thus less than fully utilized. Limiting the shear lag effect is essential for optimal development of the tubular system. A reasonable objective is to strive toward at least 75% efficiency such that the cantilever component in the overall system deflection under wind load dominates.

The framed tube in structural steel requires welding of the beam-column joint to develop rigidity and continuity. The formation of fabricated tree elements, where all welding is performed in the shop in a horizontal position, has made the steel-frame tube system more practical and efficient, as shown in Fig. 4.93. The trees are then erected by bolting the spandrel beams together at midspan near the point of inflexion.

The column spacing in steel-framed tubular buildings must be evaluated to balance the needs for higher cantilever efficiency through closer spacings with increased fabrication costs. The use of deeper, built-up sections versus rolled members is also a matter of cost-effectiveness. A survey of steel quantities for completed tubular buildings is shown in Fig. 4.94. The buildings range from 40 to 110 stories, and column spacings generally range from 3 to 4.5 m (10 to 15 ft) on center, with spacings as close as 1 m (3.28 ft) in the case of the 110-story World Trade Center twin towers, New York (Fig. 4.95). These towers are examples whereby the structuralist notion of a punched wall tube with extremely close exterior columns is architecturally exploited to express visually the inherent verticality of the high-rise building.

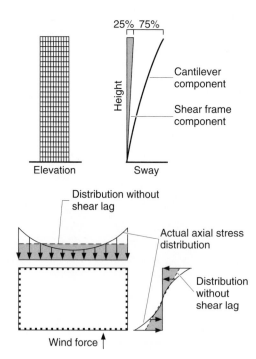

Fig. 4.92   Framed tube behavior.

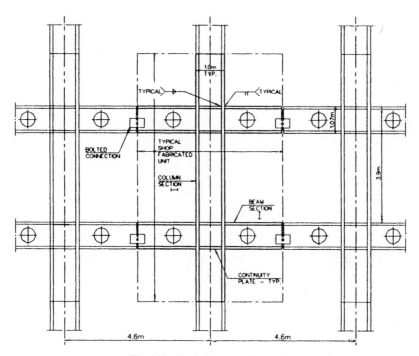

Fig. 4.93   Typical tree erection unit.

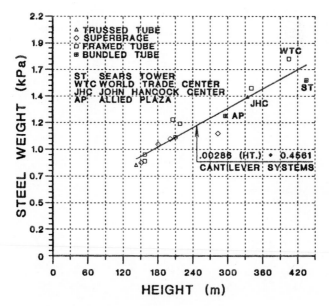

Fig. 4.94   Cantilever systems, steel quantity versus height.

3 The Trussed Tube

As the tubular concepts were being developed in the 1960s, it became apparent that there was a certain building height range for which the framed tube could be efficiently adapted. For very tall buildings, the dense grid of beam and column members has a decided impact on the facade architecture. The need to control shear lag and improve the

Fig. 4.95   World Trade Center, New York. (*Courtesy of Leslie Robertson and Assoc.*)

system efficiency can only be realized by relatively small perforations in the tubular walls. The problem becomes particularly acute at the base of the building, where architectural planning typically demands open access to the building interior from the surrounding infrastructure with as little encumbrance as possible from the exterior framework. A number of elegant solutions involving the transfer and removal of the exterior columns at the base of the building have been formulated (Figs. 4.91, 4.95, and 4.96), but characteristically include an associated material premium.

The trussed tube system represents a classic solution for a tube uniquely suited to the qualities and character of structural steel. The ideal tubular system is one which intercon-

Fig. 4.96   Two Shell Plaza, Houston, Texas. (*Courtesy of Skidmore Owings and Merrill.*)

nects all exterior columns to form a rigid box, which can resist lateral shears by axial forces in its members rather than through flexure. This is achieved by introducing a minimum number of diagonals on each facade and making the diagonals intersect at the same point at the corner column. The system is tubular in that the fascia diagonals not only form a truss in the plane but also interact with the trusses on the perpendicular faces to affect the tubular behavior. This creates the X form between corner columns on each facade.

It is clear that for buildings of this type, the structural discipline sets the basis for the overall exterior architecture (Fig. 4.97). A significant advantage of the trussed tube system is that relatively broad column spacings can be used in contrast to framed tube systems. The result is large clear spaces for windows, a particular characteristic of steel buildings.

To extend the concept of a trussed tubular system to reinforced concrete construction, a diagonal pattern of window perforations in an otherwise framed tube construction is filled in between adjacent columns and girders. The result is a reduction in shear lag for the system under wind loads. As with the steel-framed trussed tube, an additional benefit associated with the system is that the facade diagonalization serves to equalize the gravity loads in the exterior columns (Fig. 4.98). This can have a significant impact on member proportions and foundation design.

The principle of facade diagonalization can readily be used for partial tubular concepts. For example, in long rectangular buildings, the end frames along the short face may be diagonalized, whereas the long face is either a framed tube or a moment resisting frame (Fig. 4.99). The end-diagonalized frame may be in the form of a channel or C shape to provide wind resistance in both directions. The diagonalization can also vary from full cross-facade diagonals to smaller Xs, thus transforming each facade into a diagonalized braced frame form. Many variations are possible, each having an impact on the exterior architecture.

4   The Bundled Tube

In their purest form, tubular systems are generally applicable to prismatic vertical profiles. For varying profiles and buildings involving significant vertical facade offsets, the discontinuity of the tubular frame to fit the shape suffers from serious inefficiencies. The system can, however, be readily adapted to a variety of nonrectilinear, closed-plan forms, including circular, hexagonal, triangular, and other polygonal shapes. The most efficient shape is the square, whereas the triangular tube has the least inherent efficiency. The high torsional stiffness characteristic of the exterior tubular system has advantages in structuring unsymmetrical shapes.

The need for vertical planning modulation and the control of shear lag for very tall buildings led to the development of the bundled, or modular, tube concept. Figure 4.100 shows a study performed on the effect of the tube size on cantilever efficiency. Each building was structured with the same column spacing and member proportions. The cantilever efficiency was computed as the ratio of the actual top deflection to the top deflection computed using the full cantilever moment of inertia without shear lag. The results indicate that as the tube size is made smaller in relation to its height (higher aspect ratio), a significantly higher aspect ratio is obtained. This led to the notion that if the overall tubular shape is structured into small cells by the inclusion of interior tubular lines, then the overall structure would have the cantilever efficiency associated with that of a single cell with high aspect ratio. The effect of the bundling on the distribution of axial forces in the tubular columns is indicated in Fig. 4.101. The interior tubular web lines serve to reduce the effects of shear lag on the column in the middle of the windward and leeward flange frames.

Fig. 4.97 Bank of China Tower, Hong Kong. (*Courtesy of Leslie Robertson and Assoc.*)

The bundled tube concept allows for wider column spacings in the tubular walls than would be possible with only the exterior framed tube form. It is this spacing which makes it possible to place interior frame lines without seriously compromising interior space planning. In principle, any closed-form shape may be used to create the bundled form (see Fig. 4.102). The ability to modulate the cells vertically can create a powerful vocabulary for a variety of dynamic shapes. The bundled tube principle therefore offers great latitude in the architectural planning of a very tall building.

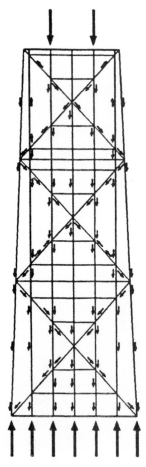

Fig. 4.98 Trussed tube, gravity load redistribution.

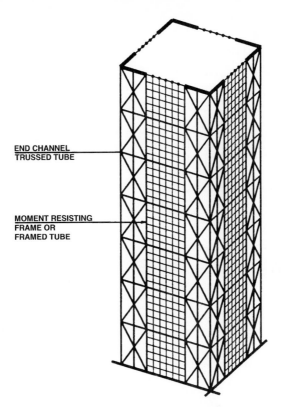

Fig. 4.99   Partial tubular system.

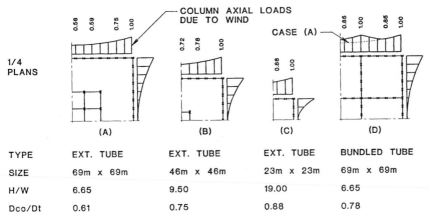

Fig. 4.100   Study of tubular efficiency.

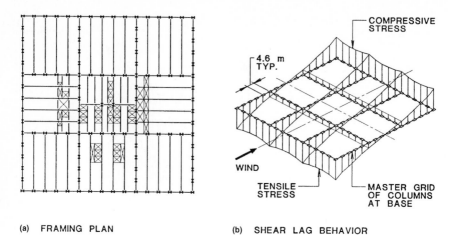

(a)   FRAMING PLAN   (b)   SHEAR LAG BEHAVIOR

Fig. 4.101   Bundled tube behavior; Sears Tower, Chicago, Illinois.

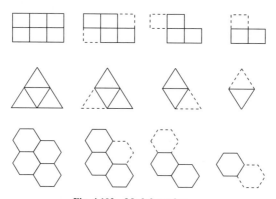

Fig. 4.102   Modular tubes.

PROJECT DESCRIPTIONS, FRAMED TUBES

Amoco Building
Chicago, Illinois, USA

Architect	Edward Durrell Stone with The Perkins and Will Partnership
Structural engineer	The Perkins and Will Partnership
Year of completion	1973
Height from street to roof	342 m (1123 ft)
Number of stories	82
Number of levels below ground	5
Building use	Office
Frame material	Structural steel
Typical floor live load	4 kPa (80 psf)
Basic wind velocity	1.4 × Chicago code
Design wind load deflection	$H/400$
Earthquake loading	Not applicable
Type of structure	Perimeter framed tube
Foundation conditions	Silty clay, sand, and gravel over massive dolomitic limestone
Footing type	Concrete caissons, 1.5 to 3.8 m (5 ft to 10 ft 3 in.) in diameter, approximately 24 m (79 ft) long
Typical floor	
Story height	3.86 m (12 ft 8 in.)
Truss span	13.7 m (45 ft)
Truss depth	965 mm (38 in.)
Truss spacing	3.05 m (10 ft)
Material	Structural steel
Slab	140-mm (5.5-in.) lightweight concrete slab; 35 MPa (5000 psi) on 38-mm (1.5-in.) steel deck
Columns	Folded plate, size not available
Spacing	3.05 m (10 ft) center to center
Material	Steel, grade 250 MPa (36 ksi)
Core	Structural steel frames carrying gravity loads only

The innovative structural concept applied to this 342-m (1123-ft)-high building resulted from the desire to achieve an efficient, simple to erect structure utilizing a perimeter tube whose behavior would closely approximate that of a pure cantilever (Fig. 4.103).

The tube comprises columns of V-shaped steel plate and deep channel-shaped bent-plate spandrel beams shop-fabricated into 3-story trees. There are 64 such columns at 3-

m (10-ft) centers around the perimeter, plus solid steel plate walls to the reentrant corners. The free inner edges of the columns are stiffened by heavy angle sections. Connections between spandrel beams comprise simple high-strength bolted joints, whereas column splices are welded at lower stories and bolted or welded at upper stories.

The floors are generally supported by 13.7-m (45-ft)-span trusses at 3-m (10-ft) centers. Trusses at successive floors attach to alternate sides of a column to effectively cre-

Fig. 4.103   Amoco Building, Chicago, Illinois. (*Photo by Jess Smith.*)

ate a concentric load in the plane of the wall. At the building corners the shorter-span diagonal girder and attached beams are wide-flange sections. The 4000 essentially iden- tical trusses and the corner beams were mass-produced in an assembly line.

Economy was achieved by creating a perimeter frame from thin steel plate spread over as much of the facade as was architecturally acceptable and by maximizing the number of geometrically identical elements. The arrangement also negated the need for subframing for the exterior curtain wall.

The space within the V-shaped columns was used for air shafts and hot and chilled water pipes for the perimeter zone. The interior zones were supplied from vertical shafts in the core.

The building contains 45,900 tonnes (50,506 tons) of steel, of which 37% is in beams and trusses and 63% in columns and reentrant corner walls. The weight of steel amounts to 161 kg/m^2 (33 psf).

181 West Madison Street
Chicago, Illinois, USA

Architect	Cesar Pelli and Associates with Shaw and Associates
Structural engineer	Cohen-Barreto-Marchertas Inc.
Year of completion	1990
Height from street to roof	207 m (680 ft)
Number of stories	50
Number of levels below ground	1
Building use	Office
Frame material	Concrete core, steel perimeter frame
Typical floor live load	2.5 kPa (50 psf)
Basic wind velocity	43 m/sec (97 mph), 100-yr return period
Maximum lateral deflection	400 mm (16 in.), 100-yr return period
Design fundamental period	8.3, 6.7 sec horizontal; 6.3 sec torsion
Design acceleration	18.4 mg peak
Design damping	1.5% serviceability
Earthquake loading	Not applicable
Type of structure	Concrete core tube with steel perimeter tube
Foundation conditions	Hardpan, 1.7-MPa (20-tsf) capacity
Footing type	Caissons, 24 m (80 ft) long, 1370 mm (4 ft 6 in.) in diameter, belled to 3-m (10-ft) diameter
Typical floor	
Story height	3.96 m (13 ft)
Beam span	10.36 m (34 ft)
Beam spacing	3.05 m (10 ft)
Beam depth	530 mm (21 in.)
Slab	140-mm (5.5-in.) composite metal deck
Columns	W350 by 745 kg/m (14 in. by 500 lb/ft)
Spacing	6.1 m (20 ft)
Material	Steel, grade A572, 350 MPa (50 ksi)
Core	Central concrete core, 62 to 28 MPa (9000 to 4000 psi)
Thickness at ground floor	400, 500, 660 mm (16, 20, 26 in.)

The 181 West Madison Street tower is a 50-story office building located at Madison and Wells Streets in the Chicago Loop (Fig. 4.104). It is a point tower, with multiple setbacks and a distinctive crown that recalls the sculpturally expressive skyscrapers of the 1920s. This is also a tower for the 1990s. It is clearly organized as a square floor plan with a center square concrete core and column-free office space (Fig. 4.105).

Fig. 4.104   181 West Madison Street, Chicago, Illinois.

181 West Madison is the tallest combination core building in Chicago. The central concrete core is surrounded by a structural steel frame and a composite floor system. The square core is 50 stories tall, for a total height of 207 m (680 ft).

The core and columns at the base of the building are supported by caissons and grade beams. Of the caissons in the project, 25% existed. Transfer-grade beams between new and existing caissons were used to take the tower's wind and gravity loads. The foundation wall on the east side of 181 West Madison required underpinning as it is a common wall with its neighbor, 10 South LaSalle Street.

Interior spans of 13.1 m (43 ft) allow a column-free interior space for maximum user flexibility. The many setbacks at the top of the building require all the perimeter columns to be transferred several times. In addition, the columns on either side of the

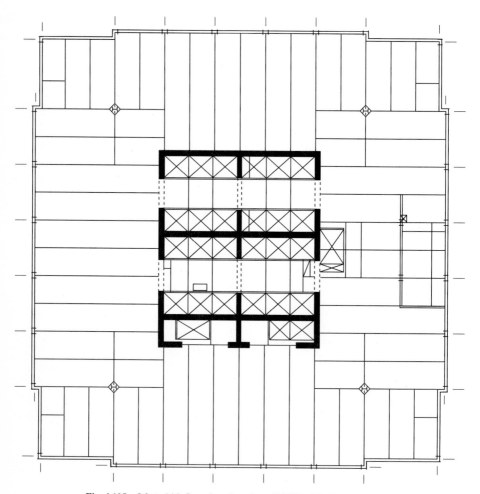

Fig. 4.105   8th to 14th floor framing plan; 181 West Madison Street.

loading dock at ground level are also transferred to increase clearance for trucks. Even with these complications, the total steel is less than 59 kg/m^2 (12 psf).

At the base of the building, on the north side, there is a four-story loggia that leads into the five-story lobby. Clad in warm white, grey, and green marble, the lobby's tall main hall has a vaulted, coffered ceiling. A bridge extends across the alley to the south to connect 181 West Madison to the Northern Trust Building.

The 102,200-m^2 (1.1 million ft^2) high-rise building was Cesar Pelli's Chicago debut, built in collaboration with the Chicago firm of Shaw and Associates.

AT&T Corporate Center
Chicago, Illinois, USA

Architect	Skidmore Owings and Merrill
Structural engineer	Skidmore Owings and Merrill
Year of completion	1989
Height from street to roof	270 m (886 ft)
Number of stories	61
Number of levels below ground	2
Building use	Office
Frame material	Composite steel-concrete perimeter frame, steel interior columns, steel floor beams
Typical floor live load	4 kPa (80 psf) levels 3 to 30; 2.5 kPa (50 psf) levels 31 to 59
Basic wind velocity	35 m/sec (78 mph), 100-yr return
Maximum lateral deflection	$H/700$
Design fundamental period	6.5 sec
Design acceleration	20 mg, 10-yr return
Design damping	1 to 1.5% serviceability
Earthquake loading	Not applicable
Type of structure	Exterior concrete-framed tube with interior gravity-load columns, trusses, and beams
Foundation conditions	18 m (60 ft) of clay over hardpan
Footing type	Belled caissons on hardpan
Typical floor	
Story height	4.0 m (13 ft 2 in.)
Truss span	14.6 m (48 ft)
Truss depth	914 mm (36 in.)
Truss spacing	4.6 m (15 ft)
Material	Steel, grade 250 and 350 MPa (36 and 50 ksi)
Slab	63-mm (2.5-in.) lightweight concrete on 76-mm (3-in.) metal deck
Columns	
Size at ground floor	1422 by 813 mm (56 by 32 in.)
Spacing	4.6 m (115 ft)
Material	Normal-weight concrete, 56 to 35 MPa (5000 to 8000 psi)
Core	Steel beams and columns for gravity load only

The AT&T Corporate Center (Fig. 4.106) consists of a 61-story office tower with rentable areas of floor plates ranging from 3250 m² (35,000 ft²) on the lowest floors to

Fig. 4.106   AT&T Corporate Center, Chicago, Illinois. (*Photo by Hedrich-Blessing.*)

1860 m² (20,000 ft²) on the highest floors (Fig. 4.107). The tower is served with five banks of passenger elevators and two service elevators. Automobile parking is provided on the two below-grade levels.

The building is clad in a combination of articulated granite panels, metal spandrel panels, and aluminum and glass curtain wall. The metal spandrel panels are painted with a detailed two-color pattern.

The structure is supported on reinforced concrete belled caissons on hardpan clay, with the bottom of the bells approximately 24 to 26 m (80 to 85 ft) below grade. A reinforced concrete slurry wall at the perimeter of the site served as the site-retention system during construction and as the permanent basement wall.

The superstructure of the building is a composite system consisting of an exterior reinforced concrete-framed tube with interior gravity steel columns and steel floor framing. The exterior concrete tube, which resists all lateral loads, has setbacks at levels 16, 29, and 44 to accommodate the articulated facade. The typical floor system consists of a 140-mm (5.5-in.) composite metal deck slab utilizing 76-mm (3-in.) cellular deck on the AT&T floors and 76-mm (3-in.) composite metal deck on the other floors. The floor framing system is a combination of built-up floor trusses and rolled beams at nominal 4.57-m (15-ft) centers. All structural steel is fireproofed with spray on cementitious material.

High-quality materials and finishes in keeping with a first-class office building were used throughout the project. The exterior was clad with a finely articulated enclosure of stone, stainless steel, aluminum, and glass. The ground- and second-floor elevator lobbies and the public spaces on levels LL-1, ground, and second have marble and granite walls and floors, with articulated ceilings and custom architectural lighting fixtures.

The structural system for the tower superstructure consists of a central core of reinforced concrete shear walls with a perimeter steel moment connected frame. Two-story outrigger trusses interconnect the concrete core and the perimeter structural steel frame

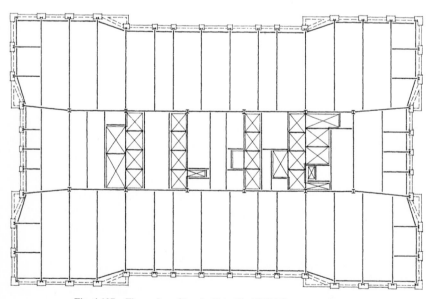

Fig. 4.107   Floor plan of levels 18 to 28; AT&T Corporate Center.

to provide a composite structure. The typical floor framing is structural steel with a 140-mm (5.5-in.) composite metal deck slab utilizing 64 mm (2.5 in.) of lightweight concrete over a 76-mm (3-in.) metal deck.

The foundation system for the tower consists of reinforced concrete, straight shaft, and rock caissons. Caissons are capped and interconnected with grade beams.

The structural system for the podium consists of a structural-steel gravity frame. The podium relies entirely on the tower for lateral load resistance. The typical floor construction of the podium is identical to that of the tower. Several steel transfer girders were used at various floor levels where column setbacks are required.

The foundation system for the podium consists of columns resting on existing belled caissons. Where column locations do not coincide with existing caisson locations, grade beams carry these column loads to the nearest caissons.

Georgia Pacific
Atlanta, Georgia, USA

Architect	Skidmore Owings and Merrill
Structural engineer	Weidlinger Associates
Year of completion	1982
Height from street to roof	230 m (753 ft 6 in.)
Number of stories	52
Number of levels below ground	2
Building use	Office
Frame material	Steel
Typical floor live load	4 kPa (80 psf)
Basic wind velocity	42 m/sec (95 mph)
Maximum lateral deflection	533 mm (21 in.), 100-yr return
Design fundamental period	6.2, 4.4 sec
Design acceleration	20 mg peak, 10-yr return
Design damping	1% serviceability; 2% ultimate
Earthquake loading	Not applicable
Type of structure	Hybrid perimeter tube—doubly folded truss on one side, beam and column frames on the other three sides
Foundation conditions	Sandy clay and silt over rock
Footing type	Bored piers to rock
Typical floor	
Story height	3.9 m (12 ft 9 in.)
Beam span	13 m (42 ft 6 in.)
Beam depth	610 mm (24 in.)
Beam spacing	3.05 m (10 ft)
Material	Steel
Slab	113-mm (4.25-in.) normal-weight concrete on metal deck
Columns	
Size at ground floor	W350 by 593 kg/m (W350 by 398 lb/ft) or built-up column
Spacing	3.05 m (10 ft)
Material	Steel, grade 350 MPa (50 ksi)
Core	
Vertical steel trusses	11.4 m (37 ft 6 in.) deep
Material	Steel, grade 350 MPa (50 ksi)
Column size	860- by 100-mm (34- by 4-in.) flanges, 368- by 100-mm (14.5- by 4-in.) web

The 52-story 96,620-m^2 (1,040,000-ft^2) Georgia Pacific center in downtown Atlanta (Fig. 4.108) has a distinctive shape with a five-sided sawtooth facade and four large setbacks facing east. The setbacks occur at floors 18, 29, 40, and 51, reflecting the reduction in service-core size as elevators are terminated progressively (Fig. 4.109). Gross floor areas vary from about 2600 to 1200 m^2 (28,000 to 13,000 ft^2).

While the combination of sawtooths and setbacks was a straightforward architectural solution, it created some extremely difficult structural engineering problems in this relatively slender building, which has a 6:1 height-to-width ratio (Figs. 4.110 and 4.111). The lateral load resisting system comprises conventional moment resisting frames on the north, south, and west facades, but such a system would have been ineffective on the east facade. Thus structural engineers Weidlinger Associates devised a unique folded truss system that extends along the full height of the building. The truss is folded along the planes of the five-sided sawtooth facade. The six vertical corner columns act as chords of the truss.

Trusses were partly shop-welded and partly field-welded, with some welded flanges almost 150 mm (6 in.) thick. Connections between columns and spandrel beams on the other three sides were also field-welded to provide for the flexibility afforded by on-site steel assembly. Field welding increased costs by 25% over shop welding, but yielded time savings of 3 months.

To achieve the required stiffness, the main elements of the bracing system operate at only about 30% maximum allowable stress levels, which resulted in large individual members, some weighing as much as 11 tonnes (11.6 tons).

Fig. 4.108   Georgia Pacific, Atlanta, Georgia.

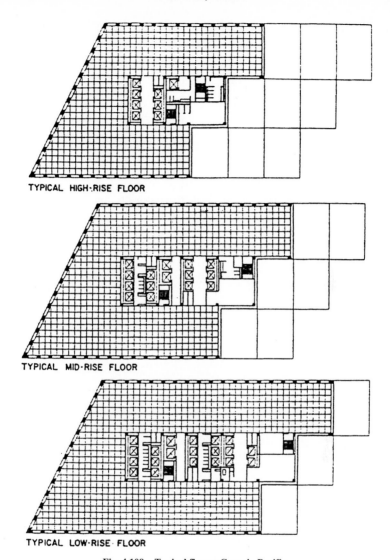

TYPICAL HIGH-RISE FLOOR

TYPICAL MID-RISE FLOOR

TYPICAL LOW-RISE FLOOR

Fig. 4.109   Typical floors; Georgia Pacific.

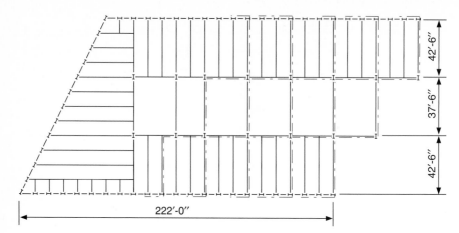

Fig. 4.110 Framing plan; Georgia Pacific.

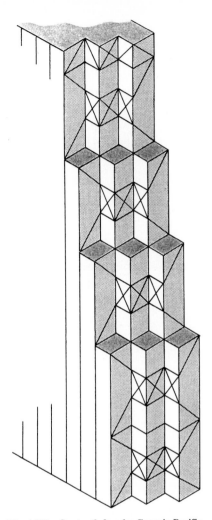

Fig. 4.111　Sawtooth facade; Georgia Pacific.

450 Lexington Avenue
New York, N.Y., USA

Architect	Skidmore Owings and Merrill
Structural engineer	Office of Irwin G. Cantor
Year of completion	1992
Height from street to roof	168 m (550 ft)
Number of stories	40
Number of levels below ground	0
Building use	Office
Frame material	Steel
Typical floor live load	2.5 kPa (50 psf)
Basic wind velocity	36 m/sec (80 mph)
Maximum lateral deflection	$H/500$, 50-yr return
Design fundamental period	5.5 sec
Design acceleration	Less than 20 mg peak
Design damping	1%
Earthquake loading	Not applicable
Type of structure	Perimeter tube with braced core
Foundation conditions	Rock, 4- to 6-MPa (40- to 60-ton/ft^2) capacity
Footing type	Piers socketed into rock
Typical floor	
Story height	3.81 m (12 ft 6 in.)
Beam span	13.4 m (44 ft)
Beam depth	460 mm (18 in.)
Beam spacing	3.05 m (10 ft)
Material	Steel, grade 350 MPa (50 ksi)
Slab	64-mm (2.5-in.) concrete over 76-mm (3-in.) metal deck
Columns	
Spacing	6.1 m (20 ft)
Material	Steel, grade 350 MPa (50 ksi)
Core	Braced steel, grade 350 MPa (50 ksi)

Placing a high-rise tower above a landmark post office structure which sits directly above a major urban rail line is a highly formidable task, which requires an unusual and innovative engineering concept. The 450 Lexington Avenue building (Fig. 4.112) is such a project and posed many challenges to the designers and contractors.

The existing landmark post office sits directly over the railroad tracks leading into New York City's Grand Central Station. The congested system of tracks made it impossible to bring the 54 tower columns down to the foundation. In addition, the track layouts totally precluded the placement of a conventional wind resisting system due to interference with train clearances. The key to the structural solution was the use of a megacolumn system. The megacolumn system forms the "legs of a table," which carries the tower's gravity and wind loads within the existing building's shell (Fig. 4.113).

The megacolumns are placed 24.4 m (80 ft) apart in the north-south direction and 51.8 m (170 ft) apart in the east-west direction. Two are 6.1 by 7.6 m (20 by 25 ft) in plan, and two are 6.1 by 2.6 m (20 by 8.5 ft). The plan sizes were governed by the avail-

Fig. 4.112   450 Lexington Avenue, New York.

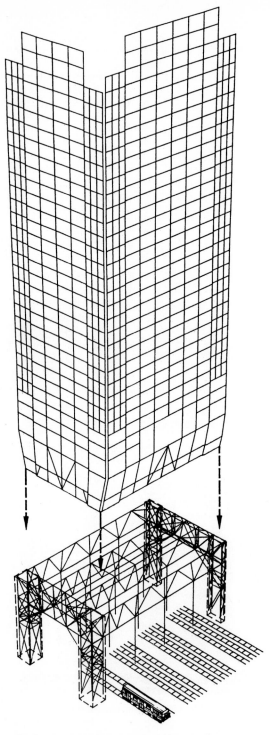

Fig. 4.113   Megacolumn system; 450 Lexington Avenue.

able space between tracks. The megacolumns are solid reinforced concrete as they rise from the foundation through the train area. At the first floor, they are composed of a steel-framed truss structure totally encased in concrete, with dimensions as noted.

Rising to the third-floor space, the megacolumns connect to massive 7.6-m (25-ft)-tall trusses. The trusses extend in both the north-south and east-west directions and connect all four megacolumns. The resulting megaframe system was referred to as the "table top."

It was the table top which picked up all the tower's columns and transferred their load to the megacolumns and to 12 strategically located conventional steel columns. The 12 intermediate columns reduce the truss spans between the megacolumns and aid in the support of gravity loads. Ultimately it was this frame which transferred all the wind loads and gravity loads to the foundations.

The flexibility of steel made it the choice material for the tower and the bulk of the megasystem. However, the concrete encasement added the needed mass and stiffness. In addition, the construction schedule and logistical constraints meant that concrete had to be the sole material within the train shed. Consequently, new crash walls located between the adjacent train tracks formed concrete wall columns. Composed of 55-MPa (8000-psi) concrete, these walls supported the intermediate columns of the mega truss system above. Utilizing concrete meant that the construction could proceed while the existing building above was still in place.

The tower's structural system is composed of a perimeter tube of columns spaced at 6.1-m (20-ft) centers. The columns are W36s and W30s for maximum efficiency. The four corners of the perimeter tube are reinforced with a vertical Vierendeel truss, which stiffens the tube significantly. Inside the core two vertical trusses are located, which rise throughout the building. These trusses supplement the perimeter tube for resistance to wind forces parallel to the shallow direction of the building.

The tower's lease space begins at the sixth floor. Below the sixth floor, all the tower's columns slope through the fifth-floor mechanical area to positions upon the top chord of the megatrusses. Figure 4.114 illustrates the thirteenth through thirty-first odd floor framing plan.

Recognition that the existing post office facility is a national landmark meant that the facade had to be maintained in its current form, whereas the central area of the existing structure was demolished to make way for the new megastructure. Consequently the facade and one adjacent bay of the structure were left in place, thereby providing the stability to the facade while demolition and construction proceeded. The remaining bay of the existing structure was known as the "doughnut" area, which was upgraded structurally and ultimately was incorporated into the final structure.

The physical complexity and intricate composite behavior of the megasystem required the use of a number of three-dimensional computer models for analysis. Lateral and vertical movement had to be determined accurately due to the impact upon the landmark facade of any excessive building motions. Consequently the lateral stiffness requirement of the megastructure was extremely stringent due to the motion constraints of the 90-year-old bunts and limestone perimeter.

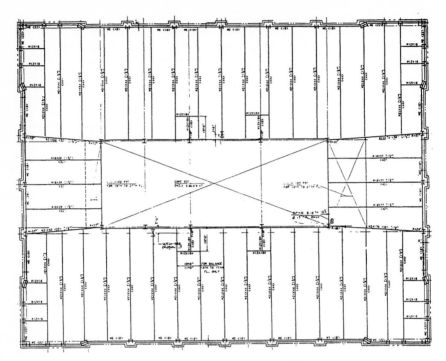

Fig. 4.114   Framing plan for odd floors 13 to 31; 450 Lexington Avenue.

Mellon Bank
Philadelphia, Pennsylvania, USA

Architect	Kohn, Pedersen, Fox
Structural engineer	Office of Irwin G. Cantor
Year of completion	1991
Height from street to roof	259 m (850 ft)
Number of stories	55
Number of levels below ground	4
Building use	Office
Frame material	Steel with composite steel-concrete columns
Typical floor live load	2.5 kPa (50 psf)
Basic wind velocity	36 m/sec (80 mph)
Maximum lateral deflection	$H/500$, 50-yr return
Design fundamental period	6 sec
Design acceleration	20 mg peak, 10-yr return
Design damping	1.5%
Earthquake loading	Not applicable
Type of structure	Perimeter tube above 6th floor, composite steel-concrete shear core below the 6th floor
Foundation conditions	6 to 9 m (20 to 30 ft) of decomposed rock over 2 to 4 MPa (20 to 40 tons) of rock
Footing type	6- to 9-m (20- to 30-ft)-deep caissons, 1800 mm (6 ft) in diameter
Typical floor	
Story height	3.8 m (12 ft 6 in.)
Beam span	13.4 m (44 ft)
Beam depth	460 mm (18 in.)
Beam spacing	3.05 m (10 ft)
Material	Steel, grade 350 MPa (50 ksi)
Slab	64-mm (2.5-in.) concrete over 76-mm (3-in.) metal deck
Columns	
Size at ground floor	W350 by 636 kg/m (W14 by 426 lb/ft) encased in 600- by 750-mm (24- by 30-in.) concrete
Spacing	2.95 m (9 ft 8 in.)
Core	Composite steel truss and concrete shear walls
Material	Steel, $F_y = 250$ MPa (36 ksi); concrete, 41 MPa (6000 psi)

As tall buildings become more slender, the dynamic behavior of the building becomes more critical. The results of the wind tunnel tests showed that the Mellon Bank tower (Fig. 4.115) had a vortex shedding problem with the cross-wind structural response being 50% larger than the response due to the code wind forces.

A comparison of various options for stiffening and damping the structure was studied and the costs of each method were estimated. It was concluded that the use of a composite structural system would be most economical. Consequently a concrete-encased

Fig. 4.115   Mellon Bank, Philadelphia, Pennsylvania.

perimeter column system coupled with a composite steel and concrete supertruss were utilized. Figure 4.116 shows the resulting floor plan.

The concrete encasement of the steel structure provided the needed damping, stiffness, and additional strength. The cost analysis performed by the construction manager proved that the composite system resulted in a more economical structure than an allsteel building. The interaction of steel and concrete and their behavior under the design loads were studied utilizing a detailed finite-element analysis.

The building's lateral system is formed by the composite perimeter columns spaced 2.95 m (9 ft 8 in.) on center, forming a perimeter tube (Fig. 4.117). Typical composite column schemes utilize the steel columns solely for erection purposes, with the bulk of the vertical load carried by the concrete. In this structure, restrictions in the overall size of the columns required the use of a truly shared composite system, with the concrete encasement and the steel columns each carrying significant portions of the vertical load.

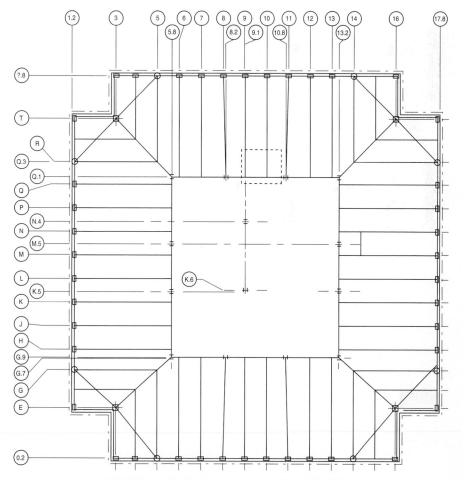

Fig. 4.116   Framing plan for floors 14 to 23; Mellon Bank.

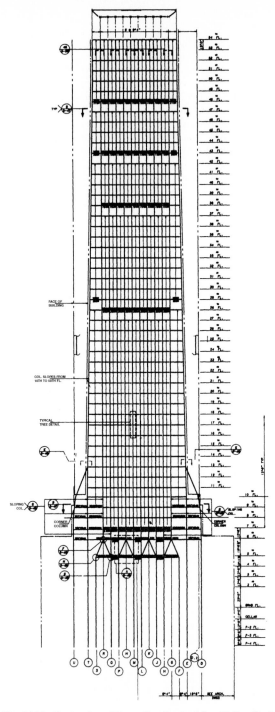

Fig. 4.117   East and west faces of perimeter tube; Mellon Bank.

Complicating the project was that none of the 52 columns in the tower continued directly to the ground. Instead, all of the perimeter columns are either sloped or picked up by trusses. The sloped column system enabled the transfer of columns into new positions, allowing for the enlargement of the lower floor plates while still maintaining column-free lease space.

Depending on the architectural constraints, groups of columns slope at different floors. The sloped columns always form a symmetrical system, whereby sloped columns on opposite sides of the floor balance out the overturning forces resulting from the slope. In numerous cases, columns are terminated upon pick-up trusses, which are also sloped to link up with their repositioned supporting columns.

A unique sloped column system occurs between the tenth and thirteenth floors, where the four inside corner columns are supported by an A-frame. Each A-frame generates significant lateral forces, which are all balanced out by again balancing one corner against the opposite corner. The floor diaphragm, being the link between all columns, plays a key role in transferring these balancing forces across the floor. The most critical diaphragms are the fifth- and sixth-floor diaphragms where, in addition to supporting most of the sloped columns, the lateral wind forces are transferred from the perimeter to the core vertical truss.

With some sloped columns generating 2000 kN (450,000 lb) in lateral force, the designer chose to place a 13.4-m (44-ft)-deep steel horizontal truss within the floor diaphragm. These trusses help transfer the wind forces to the core while passing the sloped column forces around the core to the opposite sloped column.

At the core a vertical supertruss extends from the foundation up to the sixth floor. The supertruss is constructed of steel wide-flange shapes, with the four corner columns encased in 3000- by 3000- by 600-mm (10- by 10- by 2-ft)-thick L-shaped concrete shear walls, thereby forming a composite steel and concrete supertruss. The supertruss is divided into two parts, a large 13.7-m (45-ft)-high truss between levels 6 and 3, and a single X truss on each face of the core, extending from the third level down to the foundation.

The transfer of lateral loads out of the perimeter and into the core at the sixth floor forms an optimum combination of the core and perimeter lateral systems. Transferring the wind lateral forces to the core at the sixth floor results in zero uplift forces upon the foundations.

Sumitomo Life Insurance Building
Okayama, Japan

Architect	Nikken Sekkei Ltd.
Structural engineer	Nikken Sekkei Ltd.
Year of completion	1977
Height from street to roof	75.3 m (247 ft)
Number of stories	21
Number of levels below ground	2
Building use	Office
Frame material	Structural steel from 4th floor up; concrete-encased structural steel and shear walls below 4th floor
Typical floor live load	3 kPa (60 psf)
Maximum lateral deflection	Not available
Design fundamental period	2.08 sec transverse; 2.01 sec longitudinal
Design acceleration	Level 1 EQ, 20 mg; level 2 EQ, 25 mg
Design damping	2%
Earthquake loading	$C = 0.14$
Type of structure	Structural steel perimeter tube from 4th floor up; arched concrete-encased steel frames and shear walls from ground to first floor
Foundation conditions	Gravel
Footing type	Raft
Typical floor	
Story height	3.5 m (11 ft 6 in.)
Beam span	9.9 m (32 ft 6 in.)
Beam depth	700 mm (27.5 in.)
Beam spacing	2.5 m (8 ft 2 in.)
Material	Steel, grade 400 MPa (58 ksi)
Slab	150-mm (6-in.) concrete on metal deck
Columns	
Size at 2d floor	400 by 300 mm (16 by 12 in.)
Spacing	2.5 m (8 ft 2 in.)
Material	Steel, grade 490 MPa (70 ksi)
Core	Steel frame

The main structural system of this building is a nearly square tube structure, which employs a peripheral frame in an integrated fashion (Fig. 4.118). In appearance, the tube structure has no directionality. The peripheral bearing walls of the first and second floors support the upper structure and have a large arch-shaped opening. The axial forces of the external columns of the upper tube structure are transferred by the arch-

Fig. 4.118   Sumitomo Life Insurance Building, Okayama, Japan.

shaped bearing walls of the first and second floors to the L-shaped wall columns at the four corners and thence to the foundations via bearing walls below grade.

The arch-shaped bearing walls of the first and second floors are of reinforced concrete construction with internal steel trusses (Fig. 4.119). The embedded steel structure is designed to remain elastic for long-term vertical loads and for short-term horizontal loads. The bearing walls were modeled as flat plates and analyzed by finite-element analysis. (The steel trusses were taken into consideration.) Analysis of the earthquake response was performed using a multimass model, which combined the upper tube structure with the arch-shaped bearing walls of the first and second floors. For accelerations of 3500 mm/sec^2 (11.5 ft/sec^2) during a large earthquake, the arch-shaped bearing walls remain within the allowable elastic stress range. The primary natural period in the vertical direction (considering vertical rigidity of the arch-shaped bearing walls) is 0.179 sec, so there was almost no response from the arch-shaped bearing walls due to vertical earthquake motions.

The typical floors (Fig. 4.120) are supported by 700-mm (27.5-in.)-deep trusses at 2.5-m (8-ft 2-in.) centers spanning 9.9 m (32 ft 6 in.). The spaces between the truss web members allow for the passage of ducts and pipes. The truss top chord is connected via stud shear connectors to the concrete slab. The increase in stiffness results in a frequency of vibration of the floor in excess of 9 Hz.

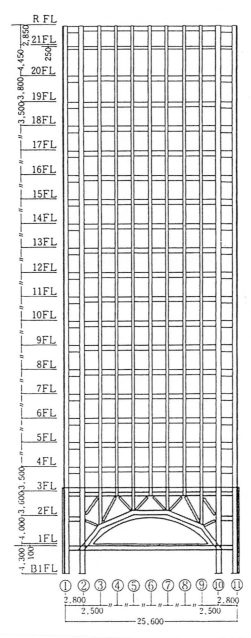

Fig. 4.119   Framework; Sumitomo Life Insurance Building.

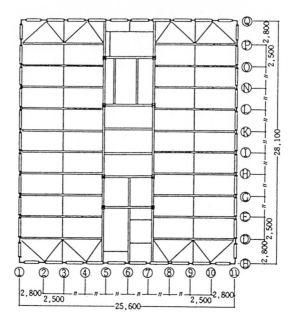

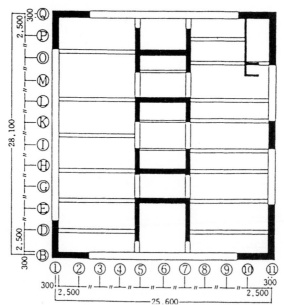

Fig. 4.120   Typical structural floor plan; Sumitomo Life Insurance Building.

Dewey Square Tower
Boston, Massachusetts, USA

Architect	Pietro Belluschi Inc. and Jung Brannen Associates Inc.
Structural engineer	Weidlinger Associates
Year of completion	1983
Height from street to roof	182 m (597 ft)
Number of stories	46
Number of levels below ground	2
Building use	Office
Frame material	Steel
Typical floor live load	2.5 kPa (50 psf)
Basic wind velocity	42 m/sec (95 mph)
Maximum lateral deflection	450 mm (18 in.), 100-yr return
Design fundamental period	5.5, 4.3 sec
Design acceleration	23 mg peak, 10-yr return
Design damping	1% serviceability; 2% ultimate
Earthquake loading	Not applicable
Type of structure	Perimeter tube
Foundation conditions	Stiff silty clay over compact glacial till
Footing type	Mat, 1800 to 2600 mm (6 to 8 ft 6 in.) thick
Typical floor	
Story height	3.81 m (12 ft 6 in.)
Beam span	9.1 m (30 ft)
Beam depth	400 mm (16 in.)
Beam spacing	2.3 m (7 ft 6 in.)
Material	Steel
Slab	133-mm (5.25-in.) lightweight concrete on metal deck
Columns	
Size at ground floor	W350 by 1088 kg/m (W14 by 730 lb/ft)
Spacing	4.57 m (15 ft)
Material	Steel, grade 350 MPa (50 ksi)
Core	Braced steel frame, grade 350 MPa (50 ksi)

After having examined many alternative systems, project designers at Weidlinger Associates concluded that a steel structure with a rigid frame around the perimeter was most economical for this 46-story building and would resolve the requirements for integrating the structure with the curtain wall (Fig. 4.121). Resistance to wind and seismic forces is provided by the framed tube forming the tower's perimeter. To economize on field work, particularly field welding, spandrel units consist of trees with columns and welded girder stubs. Field connections of the girders at the centerline between columns are bolted shear connections.

Spandrel girders on typical floors are generally 1143 mm (45 in.) deep, varying from a minimum of 900 mm (39 in.) at the top of the building to 1245 mm (49 in.) at the bottom. Columns are built-up members 760 mm (30 in.) deep along the building face, except where rolled sections are used above the thirty-third floor. Perimeter columns are

Fig. 4.121   Dewey Square Tower, Boston, Massachusetts. (*Photo by Steve Rosenthal.*)

arranged to provide open corners, that is, the ladder section always ends with a beam stub at the corner. This scheme avoided the complication of three-dimensional corner columns with welded stubs going in two directions as well as the biaxial bending problem of a corner column. Since all of the structure's lateral stiffness is provided around the perimeter, all interior beam-to-beam connections are of the simple shear type.

A variety of steels is used throughout the structure. Exterior columns and interior floor framing are of A-36 steel, girders and interior columns are A-572 grade 50, and built-up interior columns are grade 42. High-strength steels were chosen where the design was governed by strength considerations. Where the design is primarily governed by deformation criteria, as for exterior columns, lower-strength steels were used.

The tower has a structural depth of 36.57 m (120 ft) with a height-to-depth ratio of almost 5:1. This, coupled with its unusual shape, suggested the use of a wind tunnel test to verify both the magnitude and the local variations of wind forces. The wind tunnel test results very closely matched the overall forces required under the Massachusetts code. Local hot spots were found to exist particularly at the intersection between the tower and the atrium.

The analysis of the structure for lateral forces yielded information useful for future projects. It is well known that the effect of shear deformation becomes magnified with an increase in the depth-to-span ratio of the beam. Since in a frame such as this, the depth-to-span ratio is on the order of 1:5, shear deformations contribute a large part of the total lateral deformation of the structure. Specifically, in this case it was found that the lateral deflection due to drift of the building can be attributed in roughly equal parts to:

- Overall deformations of the frame (shear deflections)
- Column shortening (bending deflection)
- Shear deformation of beams and columns

Since the girder webs are relatively thin compared to the column webs, the major portion of the shear deformation is attributable to the beam web.

Wherever possible in the established program, the steel fabricator elected to substitute fillet welding for this connection between the spandrel girder flanges and the perimeter columns. This was chosen over the specified full-penetration weld.

Whenever the erection equipment would allow, the fabricator used two-story tiers for the exterior columns. These consisted of the full 7.62-m (25-ft) column, with two spandrel girder stubs on each side. The spandrel girders were then bolted together at midspan. This method kept field welding to a minimum as well as expediting the erection.

In erecting the steel tower, three self-climbing tower cranes were used in lieu of the more conventional two. This ensured maximum erection speed and facilitated the erection of the precast concrete panels, also part of the steel contractor's work. Dewey Square Tower is granite-clad on the lower two floors, with precast rain-screen panels reaching from the third floor to the sloped glass crown of the forty-sixth story. Continuous bands of tinted reflective glass alternate with bands of exposed granite aggregate set in white cement. Structural connections for the panels were developed with input from both the panel fabricator and the steel contractor. The typical panel is attached by two load-bearing connections and two lateral connections shop-welded to the perimeter columns. Floor construction consists of a 50-mm (2-in.) composite deck with an 83-mm (3.25-in.) lightweight concrete topping.

The tower starts on a concrete mat two stories below grade and rises 180 m (590 ft). The 1.8- to 2-m (6- to 8-ft)-thick concrete mat rests on hardpan, which provides an economical foundation. The area of the building surrounding the tower has columns resting on spread footings and incorporates an underdrain system below the subbasement slab.

Morton International
Chicago, Illinois, USA

Architect	Perkins and Will
Structural engineer	Perkins and Will
Year of completion	1990
Height from street to roof	170 m (560 ft) to top of clocktower
Number of stories	36 plus clocktower
Number of levels below ground	1
Building use	Office, parking, and retail
Frame material	Structural steel
Typical floor live load	2.5 kPa (50 psf)
Basic wind velocity	34 m/sec (75 mph)
Design wind load deflection	330 mm (13 in.), 50-yr return
Design fundamental period	4 sec
Design acceleration	Estimated 15 mg peak, 10-yr return
Design damping	1% serviceability
Earthquake loading	Not applicable
Type of structure	Perimeter framed tube with transfer truss at low level
Foundation conditions	Stiff clay
Footing type	Belled caissons bearing on hardpan
Typical floor	
Story height	3.81 m (12 ft 6 in.)
Beam span	12.6 m (41 ft 6 in.)
Beam depth	533 mm (21 in.)
Beam spacing	3.05 m (10 ft)
Material	Steel, grade 350 MPa (50 ksi)
Slab	140-mm (5.5-in.) lightweight concrete on steel deck
Columns	
Size at ground floor	Built-up 1640 kg/m (1100 lb/ft) max
Spacing	4.57 m (15 ft) exterior; 9.1 by 12.6 m (30 ft by 41 ft 6 in.) interior
Material	Steel, grade 350 MPa (50 ksi)
Core	Steel frames supporting gravity loads only

The Morton International building comprises a 13-story base containing commercial floors and parking for 450 cars, topped by a 23-story office tower (Fig. 4.122). The site fronts the Chicago River and contains existing railroad tracks, which had to remain fully operational during construction. Almost a quarter of the site was unable to accommodate any footings and the remainder required large spans across the tracks. Several interesting transfer systems were designed to overcome the site restraints.

The 36-story structure has typical floor spans of 12.6 m (41 ft 6 in.), but spans varying from 19.8 to 21.3 m (65 to 70 ft) were required to span the railroad tracks. This was
achieved with a series of 6-story-deep Vierendeel frames consisting of two 3.05-m (10-
ft)-deep plate girders, one at level 2 and one at level 8, connected by fully welded vertical and horizontal members. For a building of this height, a braced core would have
been the obvious means of resisting wind loads. However, in this case the railroad tracks

Fig. 4.122   Morton International, Chicago, Illinois. (*Photo by Hedrich-Blessing.*)

made this impossible and instead, a perimeter framed tube with columns at 4.57 m (15 ft) was adopted. The columns and spandrel beams were shop-fabricated into 2-story-high "ladders" with site-bolted web plate connections at midspan of the beams. This design saved 1360 tonnes (1500 tons) of steel compared to an original design with perimeter columns at 9-m (30-ft) centers.

The 13-story structure presented major challenges, which were overcome by three separate transfer structures and unusual construction requirements. Street-level concrete transfer beams 2.3 m (7 ft 6 in.) deep at 9-m (30-ft) centers span the tracks to allow a regular and efficient column setout above.

The second transfer system occurs above the roof to the southern end of the building, where no footings were able to be provided in the track zone. Trusses with major members built up from six 100- by 600-mm (4- by 24-in.) plates suspend one side of the building.

The third transfer system occurs between levels 2 and 4 and serves to redirect two rows of upper columns into one row located to avoid the tracks. The entire vertical structure above these transfer frames was erected to the roof level, and the roof top trusses were erected cantilevering beyond the floors below. This section of the building was erected 90 mm (3.5 in.) out of plumb to allow for the sway induced when the cantilevered section was erected and partially loaded.

With the roof top trusses erected, perimeter columns were suspended from the free ends of the trusses, and the floors were erected in a conventional manner from the bottom up. To equalize deflections and minimize differential movement, a load-distributing longitudinal truss was installed at level 8 between the suspended columns. This truss served a dual purpose in that it was also designed to redistribute the column load to adjacent columns should a roof-top truss fail. The roof-top trusses were provided with sufficient capacity to allow them to carry this additional load.

This challenging project received an award for Most Innovative Design of 1990 from the Structural Engineers Association of Illinois.

Nations Bank Corporate Center
Charlotte, North Carolina, USA

Architect	Cesar Pelli Associates
Structural engineer	Walter P. Moore and Associates, Inc.
Year of completion	1992
Height from street to roof	256 m (840 ft)
Number of stories	62
Number of levels below ground	2
Building use	Office, corporate headquarters, retail
Frame material	Concrete
Typical floor live load	2.5 kPa (50 psf) + 1.0-kPa (20-psf) partitions
Basic wind velocity	35 m/sec (80 mph) at 10-m (33-ft) height
Maximum lateral deflection	$H/700$, 50-yr wind
Design fundamental period	5.3 sec
Design acceleration	12 mg peak, 10-yr wind
Design damping	1.5% serviceability; 2.5% ultimate
Earthquake loading	$C = 0.53$, $Z = 0.15$, $R_w = 7.0$ intermediate moment resisting frame (IMRF)
Type of structure	Perimeter tube
Foundation conditions	Clay of variable thickness, 4.6 to 7.6 m (15 to 25 ft) over weathered bedrock
Footing type	2.4-m (8-ft)-thick core mat on weathered rock; 9- to 30-m (30- to 100-ft)-deep caissons (150 ksf), 1.5 to 1.8 m (5 to 6 ft) in diameter
Typical floor	
Story height	3.86 m (12 ft 8 in.)
Beam span	14.63 m (48 ft)
Beam depth	457 mm (18 in.) posttensioned
Beam spacing	3.05 m (10 ft)
Slab	117-mm (4.625-in.) lightweight concrete one-way, 35 MPa (5000 psi)
Columns	1370 mm (54 in.) in diameter
Spacing	6.1 m (20 ft)
Material	55 MPa (8000-psi) concrete

The Nations Bank Corporate Center is a 60-story, 256-m (840-ft) tall building in the central business district of Charlotte, North Carolina (Fig. 4.123). The building is the tallest in the southeastern United States and will dominate Charlotte's skyline into the 21st century. From a heavy stone base, the building rises with curved sides and progressive setbacks culminating in a crown of silver rods symbolizing Charlotte's nickname, "The Queen City." The exterior surface materials are reddish and beige granite and mirrored reflective glass; the granite piers narrowing at each setback. The building will serve as the corporate headquarters for Nations Bank.

A number of different feasible structural schemes were analyzed before Nations Bank and the developer together selected an economical concrete frame. A reinforced concrete frame was selected because it met both the intricate geometric requirements of the architect and the demands of the developer for economy. Shallow posttensioned concrete floors were used to span the 14.6-m (48-ft) lease depths and to achieve the desired 3.9-m (12.5-ft) floor-to-floor heights.

Fig. 4.123   Nations Bank Corporate Center, Charlotte, North Carolina.

The structural system selection followed an intensive four-phase scheme development process. This process has been used successfully in structural system selection for many other high-rise projects. The purpose of the structural scheme selection process is not only limited to finding the most economical structural system, but to finding the system that best responds to the overall building goals. Nonstructural parameters such as impact on leasing, column sizes and locations, shear wall drop-offs, construction duration, floor-to-floor heights, fire rating and integration with mechanical systems are also considered. The entire team participated in the selection process.

The selected all-concrete scheme consists of a reinforced concrete perimeter tube structure with columns spaced on 6.1 m (20 ft) centers. The perimeter frame utilizes normal weight concrete with strengths ranging from 41,300 to 55,000 kPa (6000 to 8000 psi). The external tube was selected because it was the most efficient lateral load resisting system. The tube also proved to be an economical method of dealing with the many setbacks and column transfers imposed by the building architecture. The floor system consists of a 117-mm (4.5-in.)-thick lightweight concrete slab spanning to 457-mm (18-in.)-deep post-tensioned beams. The posttensioned beams are spaced on 3 m (10 ft) centers and span as much as 14.6 m (48 ft). The 14.6-m span provides column-free lease space from the core to the perimeter. The shallow structural depth allowed the low floor-to-floor height resulting in additional savings in skin cost. Lightweight floor concrete was selected to minimize the building weight and to achieve Charlotte's unusual requirements for 3-hr fire separation. A normal weight concrete slab would have needed to be 150 mm (6 in.) thick in order to provide the fire separation, substantially increasing not only the building weight but also the floor-to-floor height.

All lateral loads are resisted by the external frame. The floor framing and core columns are sized for gravity loads. Lateral load moments imposed by compatibility of deformation with the exterior frame were found to be insignificant. The core columns were shaped to be wall-like. Column sizes ranged from 0.6 by 5.5 m (2 by 18 ft) at the lower level to 600 by 900 mm (24 ft by 35 in.) at the top of the building. The walllike colunm shapes integrated very well with the building core.

Bank One Center
Dallas, Texas, USA

Architect	John Burgee Architects with Philip Johnson
Structural engineer	The Datum/Moore Partnership
Year of completion	1987
Height from street to roof	240 m (787 ft)
Number of stories	60
Number of levels below ground	4
Building use	Office, parking
Frame material	Concrete-composite perimeter frame, steel core
Typical floor live load	2.5 kPa (50 psf) + 1.0-kPa (20-psf) partitions
Basic wind velocity	31 m/sec (70 mph) at 10-m (33-ft) height
Maximum lateral deflection	$H/500$, second order, 50-yr wind
Design fundamental period	6.8, 6.5, 3.5 sec
Design damping	2.0% serviceability; 1.5% ultimate
Earthquake loading	None
Type of structure	Perimeter tube
Foundation conditions	6.1-m (20-ft) shale and weathered limestone over unweathered limestone
Footing type	Rectangular footings on unweathered limestone; 10 MPa (100-ton/ft²) end-bearing design allowable
Typical floor	
Story height	3.84 m (12 ft 7 in.)
Beam span	14.69 m (48 ft 2 in.)
Beam depth	457-mm (18-in.)
Beam spacing	2.74 m (9 ft)
Material	Steel, A572 grade 50, 50-mm (2-in.) composite metal deck + 89-mm (3.5-in.) lightweight concrete
Columns	610-mm (2-ft)-square 100-mm (4-in.)-thick box column
Spacing	7.6 m (25 ft)
Material	Steel, A572 grade 50

Bank One Center is a postmodern tower complete with a monumental arched entry and curved roof line (Fig. 4.124). The 60-story office tower also has an atrium banking hall in its 6-story podium, semicircular arched roofs at the twenty-sixth floor and quarter-circle vaulted skylights at the fiftieth, where the shape changes from rectangular to cruciform. On top is a cross vaulted arch clad in copper and granite.

The engineering for the 148,000 m^2 (1.6 million ft^2) project is as complex as the architecture. Extensive value engineering studies were done during design development to analyze six floor framing systems and four wind framing systems. Design information for each was provided to the general contractor, who in turn studied scheduling and prices.

All four wind schemes were variations of the perimeter tube. For the early comparative design studies, Dallas building code wind forces were used. The selected scheme

Fig. 4.124   Bank One Center, Dallas, Texas.

has punched concrete walls at the building corners with infills of composite columns and steel spandrels; floors have a composite steel beam framing system.

The building's architecture requires a number of geometric changes as the structural frame rises above the below-grade levels. The cruciform shape above level 50 created two major structural problems. First, the perimeter tube had to be broken, leaving only two-dimensional rigid frames on each building facade. To control frame distortions under wind loading, two-story X-braced frames were added in the core. This required strengthened diaphragm floors to allow the transfer of wind shear forces from the frames to the perimeter tube system below. Second, corner columns at the reentrant corners of the cruciform had to be transferred to provide column-free lease space below level 50. Story-deep Vierendeel trusses spanning 13.7 m (45 ft) move these gravity column loads to the perimeter wind frame and to the core. Because of the relationship between core and perimeter columns, the trusses had to be supported at the core by two-story Vierendeel trusses spanning 8.5 m (28 ft) to the building core columns.

Central Plaza
Hong Kong

Architect	Nu Chun Man and Associates
Structural engineer	Ove Arup and Partners
Year of completion	1992
Height from street to roof	314 m (1030 ft)
Number of stories	78
Number of levels below ground	3
Building use	Office
Frame material	Reinforced concrete
Typical floor live load	3 kPa (63 psf)
Basic wind velocity	64 m/sec (144 mph), 50-yr return, 3-sec gust
Maximum lateral deflection	400 mm (15.8 in.), 50-yr return period wind
Design acceleration	Less than 10 mg, 10-yr return period (typhoon wind)
Earthquake loading	Not applicable
Type of structure	Perimeter tube and core
Foundation conditions	Fill over clay over granite bedrock; granite bedrock, 25 to 40 m (80 to 130 ft) below ground
Footing type	Machine- and hand-dug caissons to rock
Typical floor	
Story height	3.6 m (11.8 ft)
Beam span	12 m (39 ft)
Beam depth	700-mm (27.5-in.) reinforced concrete
Slab	160-mm (6.3-in.) reinforced concrete
Columns	
Size at ground floor	2-m (6.5-ft) diameter
Spacing	8.6 m (28 ft)
Material	Concrete, cube strength 60 N/mm^2 (8500 psi)
Core	Shear walls 1.3 m (4 ft 3 in.) thick at base
Material	Concrete, cube strength 60 to 40 N/mm^2 (8500 to 5800 psi)

When completed in 1992, Central Plaza was the tallest reinforced concrete building in the world (Fig. 4.125). The site is typical of a recently reclaimed area with sound bedrock lying between 25 and 40 m (80 and 130 ft) below ground level. This is overlain by decomposed rock and marine deposits, with the top 10 to 15 m (33 to 49 ft) being of fill material. A permitted bearing pressure of 5.0 MPa (56 ton/ft^2) is allowed on sound rock. The maximum water table rises to about 2 m (6.5 ft) below ground level.

Fig. 4.125   **Central Plaza, Hong Kong.** (*Courtesy of Ove Arup and Partners.*)

Wind loading is the major design criterion in Hong Kong, which is situated in an area influenced by typhoons. The Hong Kong code of practice for wind effects is based on a mean hourly wind speed of 44.3 m/sec (99 mph), 3-sec gusts of 70.5 m/sec (158 mph), and gives rise to a lateral design pressure of 4.1 kPa (82 psf) at 200 m (656 ft) above ground level.

It was clear from the outset that a multilevel basement of maximum floor area would be required. The design of a diaphragm wall, extending around the whole site perimeter, and constructed down to and grouted to rock, was completed in the first week after the site was acquired. This enabled construction to commence 3 months later (Fig. 4.126*a* to *c*).

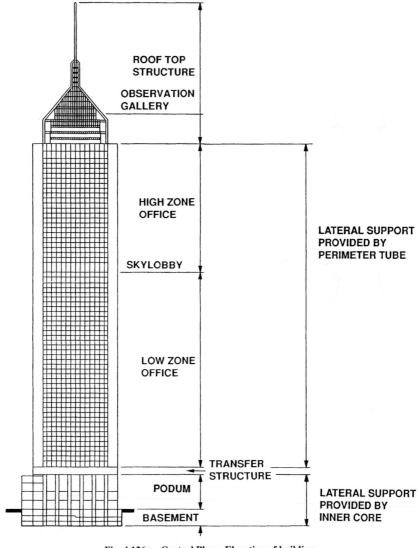

ROOF TOP
STRUCTURE

OBSERVATION
GALLERY

HIGH ZONE
OFFICE

SKYLOBBY

LATERAL SUPPORT
PROVIDED BY
PERIMETER TUBE

LOW ZONE
OFFICE

TRANSFER
STRUCTURE

PODUM

BASEMENT

LATERAL SUPPORT
PROVIDED BY
INNER CORE

Fig. 4.126*a*   Central Plaza. Elevation of building.

An initial planning assessment had indicated that up to four levels of basement could be required and the design produced catered for this. By the time construction commenced, it had been decided that only three levels would be necessary, and the construction drawings were amended accordingly.

The diaphragm wall design allowed for the basement to be constructed by the top-down method. This provided three fundamental advantages:

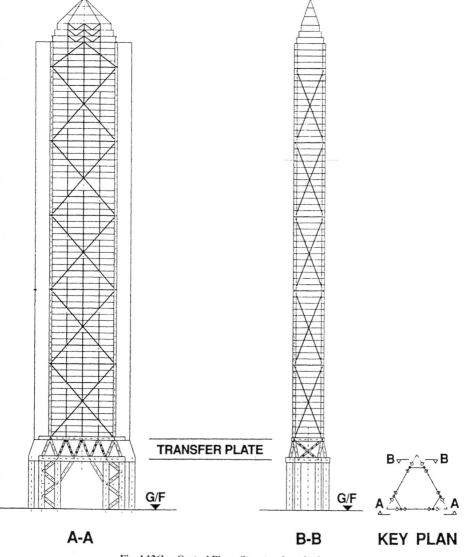

TRANSFER PLATE

A-A   **B-B**   **KEY PLAN**

Fig. 4.126b   Central Plaza. Structural steel scheme.

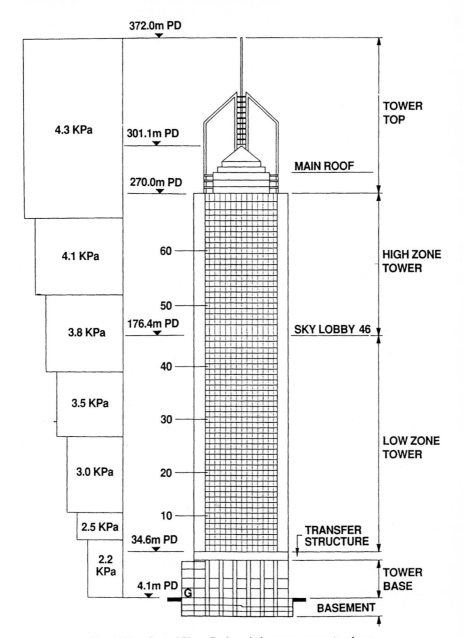

Fig. 4.126c   Central Plaza. Design wind pressure concrete scheme.

1. The superstructure can be constructed at the same time as the basement, thereby removing time-consuming basement construction from the critical path.

2. It uses the permanent floor plates for stabilizing the wall, thereby reducing likely ground movements to below those expected for open-cut excavation.

3. It created a watertight box to enable traditional hand-dug caissons to be constructed down to rock under controlled conditions.

The arrangement of the caissons was finalized once the superstructure was fixed. The maximum caisson size required was 7.4 m (24 ft) in diameter.

The original approved concept was for a 340-m (1115-ft)-tall tower with a floor plate area of approximately 1800 m² (19,370 ft²) and a corresponding slenderness ratio of more than 6. The height and slenderness of the tower necessitated the provision of an extremely stiff and strong external frame to resist the high lateral wind loads, and to limit building movement satisfactorily. After a considerable number of schematic options had been studied, it was decided that an externally cross-braced framed tube in structural steelwork best fulfilled the client's requirements. The use of structural steelwork minimized member sizes and, most importantly, resulted in savings in construction time.

The floors were conventional, with primary and secondary beams carrying metal decking with a 160-mm (6.3-in.)-thick reinforced concrete slab and a floor-to-floor height of 4 m (13.12 ft). The core was also of steelwork, designed to carry vertical load only. The core arrangement, developed by the architect with the developer's input, made it ideally suited for construction in steelwork modules (Fig. 4.127).

For the tower base, alternative options using bracing, portal action, and core bracing were developed to transfer the loads from the framed tube to the foundation level. The total weight of the steelwork for the finally adopted scheme was estimated to be on the order of 24,000 tonnes (26,900 tons), or 150 kg/m² (31 psf) of floor area. Composite action with concrete was taken into account where possible.

Subsequently a final financial review of the development proposals was carried out by the client. The construction of a diaphragm wall was under way, and while detailed design of the building was by now in progress, there was still the possibility that schematic revision could be made, provided the necessary decision could be arrived at quickly. As a result of these final studies it was decided to reduce the height of the superstructure by increasing the size of the floor plate, and to reduce the height and the complex architectural requirements of the tower base. By this means it became possible to adopt the use of a high-strength concrete solution, which limited member sizes to acceptable proportions.

Although the steelwork scheme appeared to be marginally faster, any positive advantage that would be gained could not justify the increased cost of this form of construction. Although it was always to be hidden behind the curtain walling, the developers had never been too happy about the effect of the bracing and the large corner columns on the interior spaces within the building, and the reinforced concrete scheme removed their concerns. The decision to make this major revision to the form of the building was made within a very short period of time, and detailed design of the reinforced concrete development commenced immediately to avoid delaying the construction program.

The scheme constructed has 78 stories. The height of the building is 314 m (1030 ft) above ground level, with the mast extending to a total height of 368 m (1207 ft). The floor plate area is 2214 m² (23,690 ft²).

Above the tower base, 30.5 m (100 ft) above ground level, stability is provided by the external facade frames acting as a tube. These comprise columns at 4.6-m (15-ft) centers and floor edge beams 1100 mm (43 in.) deep. The floor-to-floor height is 3.6 m

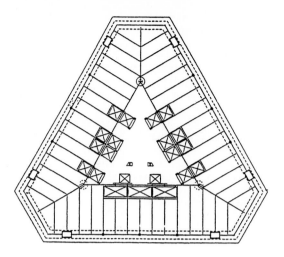

TYPICAL FLOOR PLAN – STEEL

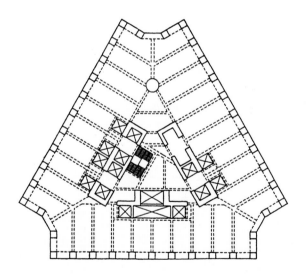

TYPICAL TOWER FLOOR PLAN – CONCRETE

(a)

Fig. 4.127   Central Plaza. (*a*) Typical office floor plans. (*b*) Foundations.

(111 ft 10 in.). The core has an arrangement similar to that of the steel scheme and, just above the tower base, it carries approximately 10% of the total wind shear.

The tower base structure edge transfer beam is 5.5 m (18 ft) deep by 2.8 m (9 ft 2 in.) wide around the perimeter. This allows alternate columns to be dropped from the facade, thereby opening up the public area at ground level. The increased column spacing, together with the elimination of spandrel beams in the tower base, results in the external frame no longer being able to carry the wind loads acting on the building. Over the height of the tower base, the core transfers all of the wind shears to the foundations. A 1-m (39-in.)-thick slab at the underside of the transfer beam transfers the total wind shear from the external frame at the inner core below.

The wind shear is taken out from the core at the lowest basement level, where it is transferred to the perimeter diaphragm walls. In order to reduce large shear reversals in the core walls in the basement and at the top of the tower base level, the floor slabs and beams are separated horizontally from the core walls at the ground floor, basement levels 1 and 2, and the fifth and sixth floors. To complete the dramatic impact of this building, the tower top incorporates a mast, which will be constructed of structural steel tubes with diameters of up to 2 m (6 ft 6 in.).

The performance of tall building structures in the strong typhoon wind climate is of particular importance. Not only must the structure be able to resist the loads in general, and the cladding system and its fixings resist higher local loads, but the building must also perform dynamically in an acceptable manner such that predicted movements lie within acceptable standards of occupant comfort criteria. To ensure that all aspects of the building's performance in strong winds will be acceptable, a detailed wind tunnel study was carried out by Professor Alan Davenport in the Boundary-Layer Wind Tunnel at the University of Western Ontario.

When completed, this project became the tallest reinforced concrete building structure in the world. For such a tall building it is not appropriate to adopt the strength of

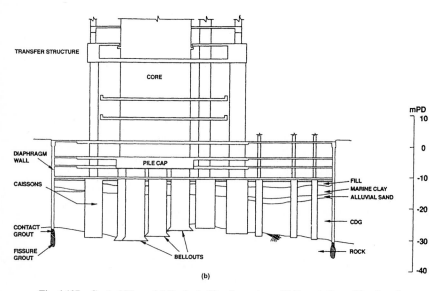

Fig. 4.127   Central Plaza. (*a*) **Typical office floor plans.** (*b*) **Foundations.** (*Continued*)

materials commonly used for normal buildings in Hong Kong. In order to reduce the size of the vertical structure it was decided to use high-strength concrete [28-day cube strength of 60 MPa (8500 psi)]. This is the first private-sector development in Hong Kong for which approval has been granted by the Hong Kong building authority for the use of such a material. Considerable research took place into materials and mix design, and many trials were carried out, including mock-ups of the large-diameter columns to check on temperature effects. As a result of this, cooling was introduced into the major pours.

The use of higher strengths was considered, but it was decided against it since it was considered by the development team that the material chosen could be produced without difficulty from materials readily available in Hong Kong.

Hopewell Centre
Hong Kong

Architect	Gordon Wu and Associates
Structural engineer	Ove Arup and Partners
Year of completion	1980
Height from street to roof	216 m (708 ft)
Number of stories	64
Number of levels below ground	1
Building use	Offices above parking and commercial podium
Frame material	Reinforced concrete
Typical floor live load	3 kPa (63 psf)
Maximum lateral deflection	150 mm (5.9 in.), 50-yr return period wind
Design acceleration	16 m*g* peak, 2-yr return period
Earthquake loading	Not applicable
Type of structure	Perimeter tube and internal core
Foundation conditions	Sound granite very close to ground level
Footing type	Pad footings on rock
Typical floor	
Story height	3.35 m (11.0 ft)
Beam span	12.3 m (40 ft)
Beam depth	686-mm (27-in.) reinforced concrete
Slab	100-mm (5.9-in.) reinforced concrete
Columns	
Size at ground floor	1.45 by 1.22 m (4.75 by 40 ft)
Spacing	3 m (10 ft)
Material	Concrete, cube strength 40 N/mm² (5800 psi)
Core	Shear walls, 762 mm (30 in.) thick at base; circular in plan
Material	Concrete, cube strength 40 N/mm² (5800 psi)

The Hopewell Centre is situated on a steeply sloping site, one entrance being at ground floor and a second main entrance to the rear of the building at the seventeenth floor (Fig. 4.128). The tower itself is founded on pad footings at levels varying between the underside of the basement and the third floor. Stability is principally provided by the perimeter tube structure formed by 48 columns at a spacing of 3 m (10 ft), linked by 1670-mm (66-in.)-deep spandrel beams at each floor level. Some assistance is also provided by the internal core. Shears are transferred to the foundations at the third-floor level through a 457-mm (19-in.)-thick floor slab (Fig. 4.129). The entire vertical structure was constructed using slip-forming techniques. The main office floors use a radial beam and slab system and were formed using fiberglass molds (Fig. 4.130). Using these techniques, construction progressed at a rate of 4 days a floor.

Fig. 4.128   Hopewell Centre, Hong Kong. (*Courtesy of Ove Arup and Partners.*)

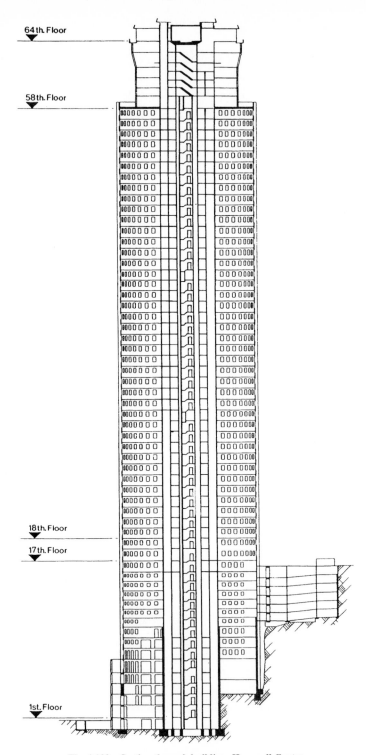

Fig. 4.129   Section through building; Hopewell Centre.

258

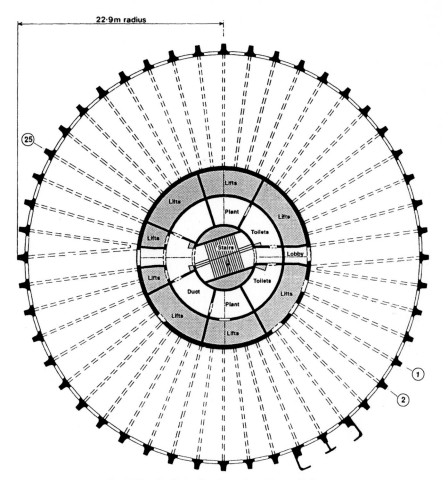

Fig. 4.130   Typical office floor plan; Hopewell Centre.

PROJECT DESCRIPTIONS, TRUSSED TUBES

First International Building
Dallas, Texas, USA

Architect	Hellmuth Obata and Kassabaum, Inc.
Structural engineer	Ellisor and Tanner, Inc.
Year of completion	1974
Height from street to roof	217 m (714 ft)
Number of stories	56
Number of levels below ground	2
Building use	Office
Frame material	Structural steel
Typical floor live load	2.5 kPa (50 psf)
Basic wind velocity	31 m/sec (70 mph)
Maximum lateral deflection	$H/500$, 50-yr return period
Earthquake loading	Not applicable
Type of structure	Trussed tube
Foundation conditions	Limestone, 4.3-MPa (40-ton/ft^2) capacity
Footing type	Spread footings
Typical floor	
Story height	3.81 m (12 ft 6 in.)
Beam span	12.27 m (40 ft 3 in.)
Beam depth	460, 530 mm (18, 21 in.)
Beam spacing	3.81, 10.97 m (12 ft 6 in., 36 ft)
Slab	83-mm (3.25-in.) lightweight concrete on 76-mm (3-in.) metal deck
Columns	
Size at ground floor	533 by 584 mm (21 by 23 in.)
Spacing	7.62 m (25 ft)
Material	Steel, grade 350 MPa (50 ksi)

The 56-story First International Building with a height of 217 m (714 ft) has 176,500 m^2 (1.9 million ft^2) of space (Fig. 4.131). There are an adjacent 13-story self-park garage and a 10-station drive-up banking facility. Tandem elevators handle the vertical movement of building occupants during peak traffic periods. Each of the 24 passenger elevator shafts has two elevator cabs, mounted one on top of the other and moving on a single set of cables.

The exterior dimensions of the office tower are 55 by 55 m (181 by 181 ft). The exterior column spacing is 7.62 m (25 ft). There is a column-free span from the core to the exterior columns of 12.27 m (40 ft 3 in.).

The design incorporates the trussed tube structural system in the exterior frame, utilizing large X braces, each covering 28 floors, two to a side. Because of the use of large X-bracing elements on the four exterior walls to resist lateral wind forces plus some

Fig. 4.131　First International Building, Dallas, Texas.

gravity loads, wind frames or trusses in the interior core are eliminated. The two X braces on each side consist of diagonal steel wide-flange members whose outside dimensions are approximately 610 by 610 mm (24 by 24 in.). The gusset plates are approximately 3 m (10 ft) wide and 3.6 m (12 ft) tall (Fig. 4.132).

Corner box columns 610 mm (24 in.) square are used in the basement and are fabricated from 152-mm (6-in.)-thick steel plates. These take the heaviest loads accumulating from the diagonal bracing of two sidewalls. The corner gusset assemblies are L-shaped in section and were welded by the electroslag process.

Another structural design concept is the stub-girder system. This minimizes structural costs through a reduction in the amount of steel required for floor framing and a lessening of the building's floor-to-floor height. The built-up girder system consists of stubs that are fabricated onto structural beams (Fig. 4.133). The wide-flange beam acts as a bottom chord whereas the short stubs act as web members. The 159-mm (6.25-in.) lightweight concrete slab functions in composite action with the steel as the top chord. The overall effect is that the slab and beam function as flanges, whereas the stubs function as web struts in a Vierendeel truss. The stub girders permit unobstructed runs of mechanical ducts without web openings in the beams. They also support the semicontinuous floor beams.

An electrified floor system was used for the first time with the stub-girder concept. Also, a longer girder is used than in previous applications of the system. In addition to a detailed computer analysis of the stub-girder design for this project, actual load tests were made to further verify the design concept.

The building was topped out in 66 weeks from groundbreaking and in 10 months from the erection of the first piece of structural steel. This project received the first Consulting Engineers Council of Texas "Eminent Conceptor Award for the Most Outstanding Engineering Project" in 1974.

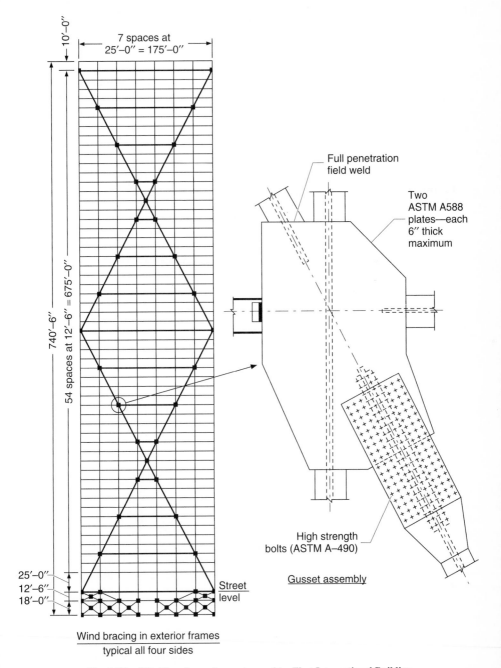

Fig. 4.132   Wind bracing and gusset assembly; First International Building.

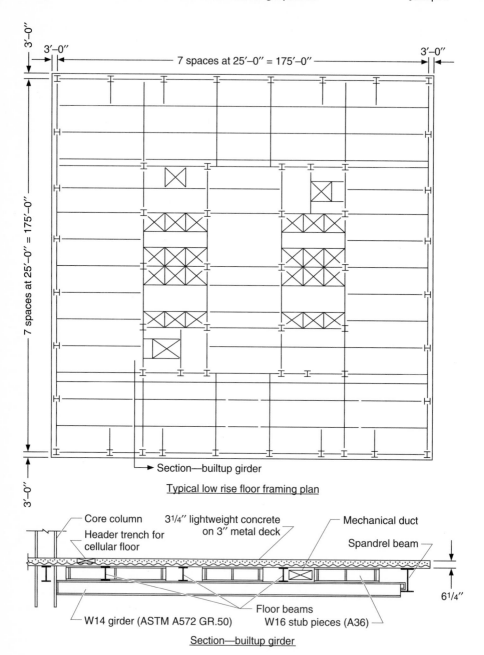

Typical low rise floor framing plan

Section—builtup girder

Fig. 4.133 Typical framing plan and built-up girders; First International Building.

Onterie Center
Chicago, Illinois, USA

Architect	Skidmore Owings and Merrill
Structural engineer	Skidmore Owings and Merrill
Year of completion	1985
Height from street to roof	174 m (570 ft)
Number of stories	57
Number of levels below ground	1
Building use	Commercial, parking, offices, apartments
Frame material	Reinforced concrete
Typical floor live load	2.5 kPa (50 psf)
Basic wind velocity	34 m/sec (75 mph)
Maximum lateral deflection	$H/500$, 100-yr return period
Design fundamental period	Not available
Design acceleration	Not available
Design damping	1 to 1.5% serviceability
Earthquake loading	Not applicable
Type of structure	Perimeter diagonally braced frames, flat-plate floors
Foundation conditions	27 m (90 ft) of clay over hardpan
Footing type	1.5-m (5-ft)-diameter caissons, belled to 3.6 m (12 ft)
Typical floor	
Story height	Apartments 2.62 m (8 ft 7 in.)
Slab	178-mm (7-in.) flat plate, spanning 6.1 by 6.7 m (20 by 22 ft)
Columns	
Size at ground floor	483 by 533 mm (19 by 21 in.)
Spacing	1.68 m (5 ft 6 in.) at perimeter
Material	49-MPa (7000-psi) reinforced concrete
Core	Not applicable

Onterie Center is a mixed-use 58-story building near the Lake Michigan shoreline in downtown Chicago (Fig. 4.134). The building has a total area of 85,000 m² (920,000 ft²), which is divided into five distinct areas by function. On the ground floor is the main public lobby and 1860 m² (20,000 ft²) of commercial space. The single-level basement and the four floors above the lobby are a parking garage. Floors 6 to 10, at the tapering base, provide office space grouped around two interior atriums. The sky lobby at level 2 includes a health club, swimming pool, hospitality room, and mechanical equipment space. The remaining floors 12 to 58 consist of 593 one-, two-, and three-bedroom apartments (Fig. 4.135).

Because mixed-use buildings need flexibility of core layout and column spacing, it was desirable to utilize only the exterior frame for the resistance of lateral loads. In the

Fig. 4.134   Onterie Center, Chicago, Illinois.

Onterie Center tower all of the lateral forces are resisted by closely spaced reinforced concrete exterior columns and spandrel beams. Additional lateral stiffness and structural efficiency were achieved by infilling window spaces with concrete in a diagonal pattern. These panels act not only as diagonal braces but as shear panels as well.

The diagonal effect of the shear panels tends to even out the gravity load on the columns and also to reduce shear lag in the tube frame under wind loading. The entire lateral load is thus resisted by two diagonally braced channels, located one at each end of the tower structure. Interior columns carry gravity loads only. The absence of a lateral load resisting core wall system allows a maximum of flexibility in planning interior space and eliminates the problem of differential axial shortening.

Three-dimensional computer modeling was used to analyze both gravity and wind load cases.

Perimeter columns are 480 by 510 mm (19 by 20 in.) at 1.68-m (5-ft 6-in.) centers. The 510-mm (20-in.)-thick infill panels contain diagonal reinforcing bars as well as horizontal and vertical bars. The concrete strength for the exterior frames and interior columns varies from 52 to 28 MPa (7500 to 4000 psi). The floors are flat slabs with thicknesses of 178 mm (7 in.) for apartments and 216 mm (8.5 in.) for commercial floors, using 35-MPa (5000-psi) concrete. Interior columns are spaced at 6.71-m (22-ft) centers. The external structural members are insulated to minimize differential-temperature induced deformations between perimeter and internal columns.

The diagonal shear panels used in the Onterie Center produce a high level of structural efficiency and create a distinctive architectural appearance. A similar system has been used on 780 Third Avenue, New York (see Fig. 4.137).

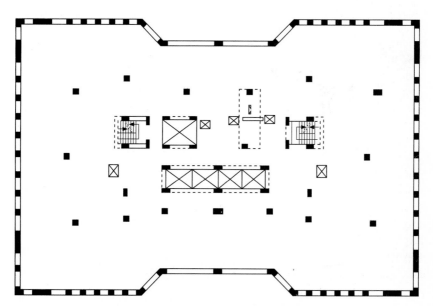

Fig. 4.135   Typical plan, 13th to 57th floor; Onterie Center.

John Hancock Center
Chicago, Illinois, USA

Architect	Skidmore Owings and Merrill
Structural engineer	Skidmore Owings and Merrill
Year of completion	1969
Height from street to roof	344 m (128 ft)
Number of stories	100
Number of levels below ground	2
Building use	3 floors commercial, 8 floors parking, 25 floors offices, 50 floors apartments
Frame material	Structural steel
Typical floor live load	2.5 kPa (50 psf)
Basic wind velocity	34 m/sec (75 mph)
Maximum lateral deflection	H/500, 100-yr return period
Design fundamental period	Not available
Design acceleration	Not available
Design damping	1% serviceability
Earthquake loading	Not applicable
Type of structure	Diagonally braced perimeter framed tube
Foundation conditions	41 m (135 ft) of clay over dolomitic limestone
Footing type	2.4-m (8-ft)-diameter caissons to rock
Typical floor	
Story height	Office 3.81 m (12 ft 6 in.); apartments 2.82 m (9 ft 3 in.)
Beam span	Office 15.2 m (50 ft); apartments 13.7 m (45 ft)
Beam depth	Office and apartments 610 mm (24 in.)
Beam spacing	Office 3.05 m (10 ft)
Material	Steel, grade 250 MPa (36 ksi)
Slab	127-mm (5-in.) concrete on metal deck
Columns	
Size at ground floor	965 by 965 mm (38 by 38 in.) built up
Spacing	12.2 m (40 ft)
Material	Steel, grade 250 MPa (36 ksi)
Core	Not applicable

The 100-story John Hancock Center is a mixed-use building with commercial, parking, office, and apartment floors (Fig. 4.136). There are approximately 93,000 m² (1 million ft²) each of office and apartment space and a combined total of 75,000 m² (800,000 ft²) of parking and commercial space. The building tapers from a ground floor approximately 48.8 by 79.2 m (160 by 260 ft) with a clear span from core to windows of 18.3 m (60 ft) to a roof of 30.5 by 48.8 m (100 by 160 ft) and a clear span of 9.15 m (30 ft).

Initial schemes envisaged twin buildings, one commercial and one residential, occupying virtually the entire site. However, environmental considerations led to the adoption of a single building, allowing setbacks from the streets and plaza space at ground level. As apartment usage requires relatively shallow depth from windows to core to provide views and natural light to all rooms whereas office usage can accept more depth, a natural consequence of these requirements could have been a tiered building. Instead, a tapered building was devised by placing the largest feasible apartment on the forty-sixth

Fig. 4.136   John Hancock Center, Chicago, Illinois.

floor (the first of the residential floors) and the largest office floor at the bottom. The taper was extended upward until all of the developer's requirements were met. The tapered form allowed a continuous structure to be used on the facade to create a tapered tube.

The structural system consists of columns and spandrel beams and diagonal cross bracing, all acting together to form an exterior tube. The requirements of the diagonals imposed a very rigid geometric discipline on the building. The diagonals from each face had to intersect at a common point on the corners so that wind shear, carried as axial loads in the web side diagonals, could be transferred directly to the flange side diagonals. The diagonal X bracing is continuous from face to face and is connected to the columns, allowing load to be transferred from bracing to columns and vice versa. The beams are provided at the levels where diagonals intersect corner columns so that the diagonals could redistribute the gravity load among the columns. The gravity load in the diagonals causes them to always be in compression under wind load, leading to much simplified connections. The redistribution of gravity load also allowed all columns on each face to be made equal in size.

A typical tier of the tube consists of a primary system comprising columns, diagonals, and spandrel beam ties at levels where the diagonals intersect columns at a floor level, and a secondary system comprising the spandrel beams at other levels. The primary structure was required to develop continuity and to transmit axial loads. The lateral load is resisted 80% by cantilever action and 20% by frame action. This is due to the diagonals creating an almost uniform column load distribution across the flange face; there is very little shear lag. The structural efficiency is demonstrated by a steel weight of only 145 kg/m^2 (29.7 psf).

The floors are a composite system of steel beams and a 127-mm (5-in.) semilightweight slab. On apartment levels the beams are arranged in such a way that they align with partitions and the soffit of the slab is plastered and used as the finished ceiling. The geometric discipline of the exterior diagonal module is maintained by three typical office story heights equaling four typical apartment story heights.

To achieve simple joints, the columns, diagonals, and ties are all fabricated I sections. The thickest plate is 152 mm (6 in.) and the largest column is 915 by 915 mm (36 by 36 in.). Interior columns were designed for gravity load only, using rolled and built-up sections. A36 steel was used for nearly all members.

Joints consist of double gusset plates to which diagonal members are connected by grade A490 bolts. Spandrel ties are field-welded to columns above and below, similar to typical column splices with bolted webs and partial-penetration flange welds. All gusset plate assemblies were shop-welded with corner gusset plate assemblies requiring stress relief.

The simple detailing resulted in an erection rate of three floors per week.

780 Third Avenue
New York, N.Y., USA

Architect	Skidmore Owings and Merrill
Structural engineer	Robert Rosenwasser Associates
Year of completion	1983
Height from street to roof	174 m (570 ft)
Number of stories	50
Number of levels below ground	2
Building use	Office
Frame material	Concrete
Typical floor live load	2.5 kPa (50 psf)
Wind load	New York City code, 1 to 1.5 kPa (20 to 30 psf)
Maximum lateral deflection	180 mm (7 in.) at design load
Design fundamental period	4.8 sec E-W; 2 sec N-S
Design acceleration	12 mg peak, 10-yr return period
Design damping	1% serviceability; 2% ultimate
Earthquake loading	Not applicable
Type of structure	Diagonally braced exterior tube
Foundation conditions	Rock, 4-MPa (40-ton/ft^2) capacity
Footing type	Spread footings
Typical floor	
Story height	3.5 m (11 ft 6 in.)
Spandrel beams	380 mm (15 in.) deep
Slab	380-mm (15-in.)-deep one-way joist and two-way waffle slab
Material	Concrete, 31 and 28 MPa (4500 and 4000 psi)
Columns	1220 by 610 mm (48 by 24 in.) at ground floor
Material	Concrete, 41, 34, 28 MPa (6000, 5000, 4000 psi)
Core	Concrete walls and columns; concrete strength as columns

The trend toward very high-rise construction in concrete has received a big boost due to the adaptation of the first diagonally braced tube system to concrete structures. The first of its kind is the 50-story office building located at 780 Third Avenue, New York (Fig. 4.137), which was completed in March, 1983. Its very slender aspect ratio of over 8:1 is what suited it to this design approach.

The building contains close to 46,500 m^2 (500,000 ft^2) of office space. Its structural system is a hybrid, utilizing three varied systems—a truss, a tube, and, to a minor extent, frame and shear wall interaction of its remaining structural components. All systems interact to provide gravity and lateral load-carrying capacity at an efficiency not previously available. This hybrid system appears to remove any practical height limit from design in reinforced concrete (Fig. 4.138).

The "concrete tube" consists of closely spaced perimeter columns which are connected at each floor level by spandrel beams. In addition, the tube is braced by a diagonal pattern of rectangular panels, in place of window openings, between adjacent columns and girders.

Fig. 4.137   780 Third Avenue, New York. (*Courtesy of Robert Rosenwasser Assoc.*)

The building is 38 by 21 m (125 by 70 ft) in plan, with an overall height of 174 m (570 ft), consisting of a 4.4-m (14.5-ft)-high first story and 48 3.5-m (11.5-ft)-high standard stories. Perimeter columns are 1.2 m (4 ft) wide, with window openings 1.6 m (5.3 ft) wide. The column thickness reduces from 610 to 457 to 406 to 356 mm (24 to 18 to 16 to 14 in.) at floors 2, 20, and 32.

The spandrel beams, which are the solid edges of the floor construction and are flush bottom with the one-way and two-way joists, are 380 mm (15 in.) deep by 1 m (39 in.) wide, except for those at the second floor, which are 762 mm (30 in.) deep by 610 mm (24 in.) wide.

The concrete bracing panels are of the same thickness as the adjacent columns and are placed integrally with them. The purpose of adding bracing to the tube is to reduce shear lag effects, and hence improve the performance of the structure for both gravity and wind loading. The wide faces of the building have double diagonal bracing, whereas the narrow faces have only single diagonal bracing in a zigzag pattern.

The concrete strength of the columns and panels varies along the building height. The maximum strength of 41 MPa (6000 psi) is reduced to 35 MPa (5000 psi) in the middle third and to 28 MPa (4000 psi) in the top third of the structure. The concrete strength of the floor members matched 31 MPa (4500 psi) with 41-MPa (6000-psi) columns and 28 MPa (4000 psi) with the lesser-strength columns.

Another structural element in the building is the set of elevator core walls. Because of their small size and central location they are considered to be of secondary importance in their influence on the braced tube's behavior.

The wind pressure applied to the building is in accordance with the New York City building code, increasing with height in steps up to a maximum of 1.44 kPa (30 psf) at the 91.4-m (300-ft) level and above. The results of a wind tunnel aeroelastic test verified that the code's wind-pressure requirements for the design of the structure frame

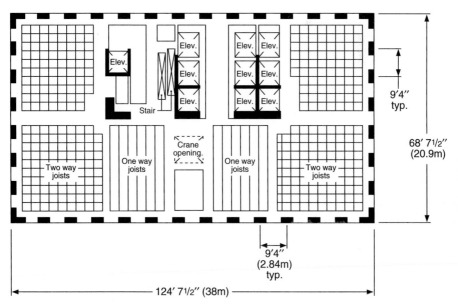

Fig. 4.138 Typical framing plan; 780 Third Avenue.

were not exceeded. The cladding design requirements were, however, upgraded on the basis of the wind tunnel test results. The projected 10-year return maximum accelerations of 12 mg registered well within the accepted industry limits for office structures.

Results from the analyses performed for 780 Third Avenue that are of particular interest are those that indicate increased cracking and reduction in the effects of shear lag by the bracing on the column forces of an unbraced tube structure.

Results of sensitivity studies and the influence of the panels on lateral stiffness are illustrated by the deflection curves in Fig. 4.139. Evidently cracking in floor members is very detrimental to the stiffness of unbraced tube structures (curves I and II), but of only secondary importance in braced tubes (curves III and IV). The stiffening effect of the bracing is demonstrated both in the reduced sway and in the modified-mode shape of the deflection curve (curve I versus curve III). The unbraced tube deflects in a wall-frame configuration, with concavity downwind in the lower part, concavity upwind in the upper part, and a point of contraflexure at about two-thirds of the height. The braced tube deflects in a more strongly flexural shape with a much higher point of contraflexure. The component of the tube's deflection due to racking shear of the columns and

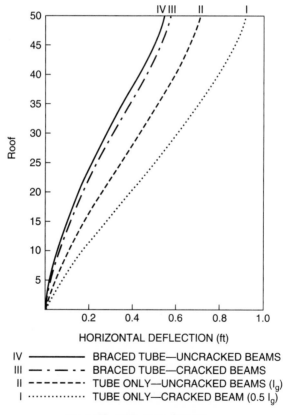

IV ——————— BRACED TUBE—UNCRACKED BEAMS
III — · — · · BRACED TUBE—CRACKED BEAMS
II — — — — — · TUBE ONLY—UNCRACKED BEAMS (I_g)
I ············· TUBE ONLY—CRACKED BEAM (0.5 I_g)

Fig. 4.139   Deflections of structure.

spandrels was, therefore, reduced significantly by the bracing. This is further supported by the small increase in the overall deflection when the spandrel stiffnesses are assigned the large (50%) reduction to account for cracking.

The deflection curve for the braced structure with cracked beams shows an increase in drift of 4% at the top, and a minimum increase of approximately 7% at about mid-height. The maximum drift per story, however, which occurs in the middle region of the building, was hardly affected.

The small influence on the overall lateral stiffness of the braced structure of a 50% variation in the moment of inertia of the spandrel beams indicates that their flexural stiffness, and therefore their depth, in the braced tube structure are of secondary importance. Their primary role is to act as ties or struts in developing the axial forces in the intermediate columns.

Figure 4.140 indicates the placement of the panel reinforcing. The column and spandrel beam reinforcing was extended through the panel, which was also reinforced with light orthogonal reinforcements to minimize the size of accidental cracks. Collector reinforcing, supplementing the spandrel reinforcements, was added to the top and bottom of the panel to augment the tensile requirements at the intersections. Splices in the main spandrel reinforcements were staggered to provide for tensile forces in the spandrel beams.

The construction of the concrete structure, from first footing to roof level, took 13 months to complete. The building required 16,000 m³ (21,000 yd³) of concrete and 2180 tonnes (2400 tons) of reinforcing bars. A 3-day construction cycle was easily maintained for the typical floors (a 2-day cycle would have been possible with overtime).

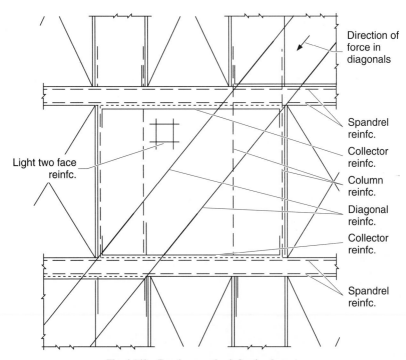

Fig. 4.140   Bracing panel reinforcing layout.

Hotel de las Artes
Barcelona, Spain

Architect	Skidmore Owings and Merrill
Structural engineer	Skidmore Owings and Merrill
Year of completion	1992
Height from street to roof	137 m (450 ft)
Number of stories	43
Number of levels below ground	1
Building use	Hotel
Frame material	Structural steel
Typical floor live load	2.87 kPa (60 psf)
Basic wind velocity	40 m/sec (90 mph) at 30 m (98 ft)
Maximum lateral deflection	$H/500$, 50-yr return period
Design fundamental period	5.2 sec
Design acceleration	Not applicable
Design damping	1% serviceability
Earthquake loading	Not applicable
Type of structure	Diagonally braced tube in the form of mega portal frames
Foundation conditions	Dense sand
Footing type	Augered straight shaft piles constructed under bentonite slurry
Typical floor	
Story height	3.00 m (9 ft 10 in.)
Beam span	Office 9.2 m (30 ft)
Beam depth	Office 457 mm (18 in.)
Beam spacing	Office 4.6 m (15 ft)
Material	Steel, A572, grade 50
Slab	75-mm composite metal deck + 60-mm (2.4-in.) concrete + 55-mm (2.1 in.) second-pour concrete
Columns	
Size at ground floor	W350 by 500 lb/ft interior; WTM 21 exterior
Spacing	9.2, 13.8 m (30, 45 ft)
Material	A572 grade 50
Core	Braced to span between mega bracing panel points; steel-braced frames in orthogonal directions

The Hotel de las Artes tower is the most prominent part of a multiuse complex in Barcelona, Spain, consisting of 5-star luxury hotel/apartment units, commercial office space, retail, parking, and health club facilities (Figs. 4.141 and 4.142). The project is

Fig. 4.141   Hotel de las Artes, Barcelona, Spain.

located along Barcelona harbor, overlooking the Mediterranean Sea, and was completed in time for the 1992 Summer Olympic Games. The Hotel de las Artes is part of an over-all plan to provide new infrastructure and private development of individual building parcels in the Olympic Village area. The tower is envisioned as one of the focal points in the reawakening of Barcelona as a major European capital.

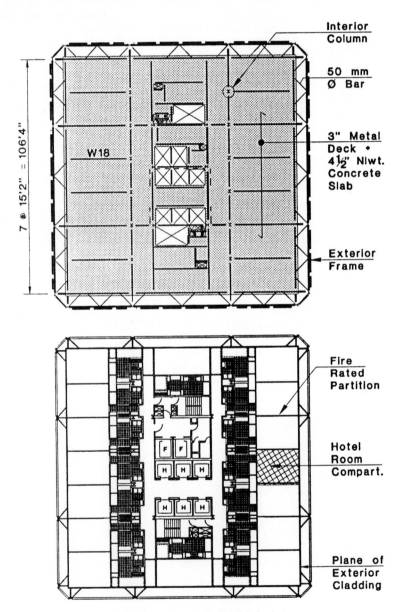

Fig. 4.142   Framework; Hotel de las Artes.

Continuing a long tradition at Skidmore Owings and Merrill, the architectural form, expression, and articulation of the tower are all based on the beauty and essence of the exposed, painted structural steel frame. The architecturally exposed X-braced frames located on the building periphery are organized on a 4-story [12-m (39-ft)] module. These frames form a fully three-dimensional system resisting all wind and seismic lateral forces as well as a portion of the tower gravity load. As the full building inertia is utilized, a very efficient lateral load resisting system is obtained, with very little steel weight premium above that required to resist the tower gravity load.

From the architectural point of view, a clear articulation of the exterior structure was desired, which is characterized by the crisp proportions of steel I beams, columns, and built-up members, as well as the honest expression of the connecting joints, both bolted and welded. The exterior curtain wall is set back 1.5 m (5 ft) from the perimeter, thereby providing a clear architectural expression of the exposed X-braced steel frame. An open, weblike structure allowing the play of daylight through the frame, much desired by the architectural design team, was balanced by the need for robustness and structural integrity, particularly at the member joints. Exterior frame members were chosen on the basis of erectability, connection detailing, accessibility for steel painting and future maintenance, and visual considerations related to the architectural aesthetic.

The issues of corrosion and fire protection were addressed in engineering the exterior exposed steel frame. Corrosion protection for the exposed steel members is provided by a durable fluorocarbon paint system designed for long life under the coastal marine environment, consisting of a shop-applied primer, undercoat, and finish coat, with a second finish coat applied in the field after erection of the steel frame. The non-fireproofed exterior structure was analyzed using the latest state-of-the-art fire engineering methods developed in Europe and the United States. Analytical methods to determine the steel temperatures as well as the character and nature of a number of hypothetical design fire events were studied. High-temperature structural analysis of the entire building frame completed the fire engineering design.

A simple, straightforward architectural composition expressing the inherent function of the structural frame, the Hotel de las Artes tower represents a prominent work combining architecture and structural engineering, marking a major international celebration in Barcelona during the summer of 1992.

PROJECT DESCRIPTIONS, BUNDLED TUBES

Sears Tower
Chicago, Illinois, USA

Architect	Skidmore Owings and Merrill
Structural engineer	Skidmore Owings and Merrill
Year of completion	1974
Height from street to roof	443 m (1454 ft)
Number of stories	110
Number of levels below ground	3
Building use	Office
Frame material	Structural steel
Typical floor live load	2.5 kPa (50 psf)
Basic wind velocity	34 m/sec (75 mph)
Maximum lateral deflection	$H/550$, 100-yr return period
Design fundamental period	7.8 sec
Design acceleration	20 mg peak, 10-yr return period
Design damping	1.25% serviceability
Earthquake loading	Not applicable
Type of structure	Bundled framed tubes
Foundation conditions	18-m (20-ft)-deep steel-lined concrete caissons
Footing type	Raft
Typical floor	
Story height	3.92 m (12 ft 10.5 in.)
Truss span	22.9 m (75 ft)
Truss depth	1016 mm (40 in.)
Truss spacing	4.6 m (15 ft)
Material	Steel, grade 250 MPa (36 ksi)
Slab	63-mm (2.5-in.) lightweight concrete on 76-mm (3-in.) metal deck
Columns	
Size at ground floor	990 by 610 mm (39 by 24 in.) built up
Spacing	4.6 m (15 ft)
Material	Steel, grade 350 MPa (50 ksi)
Core	Not applicable

The Sears Tower is the world's tallest office building with a height of 443 m (1454 ft) above ground (Fig. 4.143). It contains 362,000 m^2 (3.9 million ft^2) of office space in 109 stories.

The setbacks in the facade result from reducing floor areas required by tenancy considerations. Sears, Roebuck and Company required large floors for their operations, whereas smaller floors were best for rental purposes. The adopted bundled tube concept

Fig. 4.143   Sears Tower, Chicago, Illinois. (*Courtesy of Skidmore Owings and Merrill.*)

provided an organization of modular areas which could be terminated at various levels to create floors of different shapes and sizes (Fig. 4.144). Each tube is 22.9 m (75 ft) square, and nine such tubes make up a typical lower floor for an overall floor dimension of 68.6 m (225 ft). This square plan shape extends to the fiftieth floor, where the first tube terminations occur. Other terminations occur at floors 66 and 90, creating floor areas of 3800 to 1100 m² (41,000 to 12,000 ft²).

The structure acts as a vertical cantilever fixed at the base to resist wind loads. Nine square tubes of varying heights are bundled together to create the larger overall tube. Each tube comprises columns at 4.58-m (15-ft) centers connected by stiff beams. Two adjacent tubes share one set of columns and beams. All column-to-beam connections are fully welded. At three levels, the tubes incorporate trusses, provided to make the axial column loads more uniform where tube drop-offs occur. These trusses occur below floors 66 and 90 and between floors 29 and 31.

The two interior frames connect opposing facade frames at two intermediate points, thereby reducing the shear lag effect in the flange frames. This reduces the premium for height considerably, as shown by the relatively low unit structural steel quantity of 161 kg/m² (33 psf). The wind-induced sway is about 7.6 mm (0.3 in.) per story, and the fundamental period is 7.8 sec.

The 22.9-m (75-ft)-square floor areas of each tube are framed by one-way trusses spanning 22.9 m (75 ft) at 4.58-m (15-ft) centers. Each truss connects directly to a column with a high-strength friction-grip bolted shear connection. The span direction of these trusses was alternated every six stories to equalize gravity loading on the columns. The trusses are 1020 mm (40 in.) deep and utilize all of the available depth in the space between the ceiling and the floor slab above. The spaces between the diagonal truss web members allow the passage of up to 530-mm (21-in.)-diameter air-conditioning ducts.

Beams and columns are built-up I sections of 1070- and 990-mm (42- and 39-in.) depth, respectively. Column flanges vary from 609 by 102 mm (24 by 4 in.) at the bottom to 305 by 19 mm (12 by 2.75 in.) at the top, and beam flanges from 406 by 70 mm (16 by 2.75 in.) to 254 by 25 mm (10 by 1 in.). A total of 69,000 tonnes (76,000 tons) of structural steel was used in the project, consisting of grades A588, A572, and A36.

The steel-tube structure was shop-fabricated into units of two-story-high columns and half-span beams each side, typically weighing 14 tonnes (15 tons). The shop fabrication eliminated 95% of field welding. Automated electroslag welding was used for the butt welds of beams to columns. The continuity plates across columns at the joints were fillet-welded by the innershield process.

Because site storage space was unavailable, the frame units were delivered exactly when needed and lifted off the truck into place. Except for column splices, all field connections were grade A490 high-strength friction-grip bolts in shear connections. Exterior columns were insulated to limit the average temperature differential between these columns and interior columns.

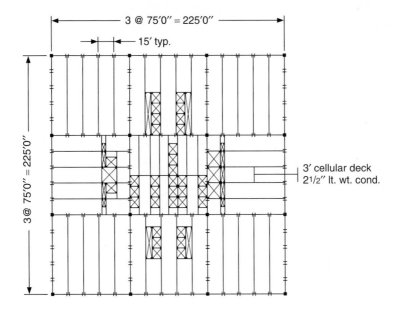

Typical framing plan (levels 1 to 50)

(a)

Plan shapes

Modular floor configuration

(b)

Fig. 4.144   Sears Tower.

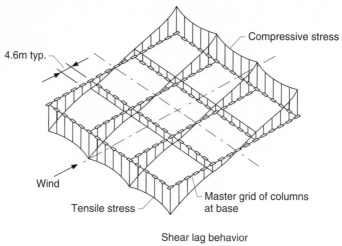

Shear lag behavior

(*c*)

Fig. 4.144   Sears Tower. (*Continued*)

Rialto Building
Melbourne, Australia

Architect	Gerard de Preu/Perrott Lyon Mathieson Pty. Ltd.
Structural engineer	Meinhardt Australia Pty. Ltd.
Year of completion	1985
Height from street to roof	243 m (797 ft)
Number of stories	63
Number of levels below ground	2
Building use	Office
Frame material	Concrete
Typical floor live load	4 kPa (80 psf)
Basic wind velocity	39 m/sec (87 mph), 50-yr return
Maximum lateral deflection	230 mm (9 in.), 50-yr return
Design fundamental period	6.1 sec
Design damping	3% serviceability; 5% ultimate
Earthquake loading	Not applicable
Type of structure	Concrete core with concrete perimeter frames
Foundation conditions	Basalt over sands and clays over mudstone
Footing type	Caissons 1500 or 1800 mm (5 or 6 ft) in diameter, 18 m (59 ft) long, socketed into rock
Typical floor	
Story height	3.9 m (12 ft 9.5 in.)
Beam span	10.5 m (34 ft 6 in.)
Beam depth	500 mm (20 in.)
Beam spacing	5 m (16 ft 5 in.)
Slab	120-mm (4.75-in.) lightweight concrete
Columns	
Size at ground floor	1.2 m (4 ft) octagonal
Spacing	5 m (16 ft 5 in.)
Material	Concrete, 60 MPa (8500 psi)
Core	Shear walls, 750 mm (30 in.) maximum thick at ground floor
Material	Concrete, 60 MPa (8500 psi)

A number of structural systems for the Rialto Building (Fig. 4.145) were initially investigated and a reinforced concrete structural system was finally adopted, with speed of construction being a prime consideration in the development of formwork and reinforcement details.

Fig. 4.145   Rialto Building, Melbourne, Australia.

The external frame of columns and beams, while being designed for the direct dead and live loads applicable, acts as an external tube in resisting lateral load. Although the plan shape is unsymmetrical and the columns are 5 m (16.4 ft) apart, analysis of the load transfer around the corners indicated reasonable three-dimensional action. The corner beams connecting the end columns are most necessary for this action. The tube effect also provides for some lateral distribution of load from the more heavily loaded columns (Fig. 4.146).

The service cores, being the major elements in the structure, were the subject of a number of detailed considerations. No sizable penetrations or rebates were permitted in the main walls. Sizing of the walls was not only for loading considerations, but was the subject of shrinkage and creep estimations and refinement for building performance. Final checking of the interacting cores and external frames was carried out using a three-dimensional finite-element analysis.

Design wind loads in the building were calculated using meteorological data available. The building is of such a height, size, and slenderness that the different approach velocities and wind directions were significant in the design. Wind tunnel tests determined design pressures for both the building and the facade. From the north, east, and west, terrain category 4 was applicable, while from the south, with Port Phillip Bay being 3 km (2 mi) distant, terrain category 1 was considered above level 30.

The lateral projection of the building, being asymmetrical, induces a wind force on the structure that does not always conform with the center of stiffness. The perimeter beams and cores have been modified to align the two centroids as closely as possible at all levels; however, a section of the building between levels 24 and 40 is subject to a twisting force. The calculated drift at the top of the tower under maximum design wind forces and incorporating this twisting is 230 mm (9 in.).

A major consideration addressed and resolved early in the design phase was the aspect of shrinkage and creep of the concrete structure. Most buildings of this size worldwide are steel framed and not subject to these types of movements.

An assessment of the creep and shrinkage of vertical elements in the project was carried out making use of research data available from the United States. Using average values derived for material properties and predictions with regard to weather and building program, a computer program was developed taking into account member size, concrete strength, reinforcement ratio, age at loading, humidity, loading conditions, and creep and shrinkage development. It was anticipated that the total nonelastic shortening of the 65-story tower would be on order of 150 to 200 mm (6 to 8 in.). Provided allowances are made in the attachment of non-load-bearing elements such as lift rails and the facade, the magnitude of this nonelastic deformation is not significant. However, differences in the magnitude of shrinkage and creep *within* a tall concrete structure is a major subject of concern, and this is particularly relevant in the case of the Rialto towers.

Long-term differential shortening between the central core and perimeter columns at the top of a typical tower building can be readily catered for as the distances between these elements are usually large. The combined shrinkage and creep to be expected after construction of the upper levels of the Rialto towers indicated differential values of 10 mm ($\frac{3}{8}$ in.) in the case of tower B and 12 mm ($\frac{1}{2}$ in.) in the case of tower A. The minimum spans involved are 9.7 m (32 ft) and 7.0 m (23 ft), respectively. However, as towers A and B form an integrated structure, a differential value on the order of 38 mm (1.5 in.) could be expected between adjacent columns at level 41 (tower B roof) due to effects of the additional 17 levels of tower A. The distance between these columns is only 4 m (13 ft), and clearly such movements cannot be tolerated in a construction of this nature. Jointing of the towers was not acceptable, and the provision of a "belt" at this level was unsuitable to the architecture, as well as inducing a long-term out-of-plumb of the top of tower A.

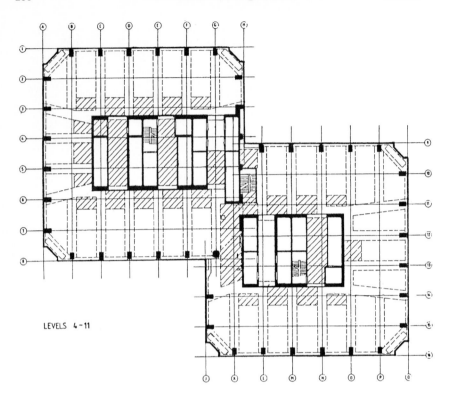

LEVELS 4-11

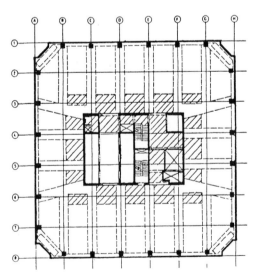

Fig. 4.146   Floor plans; Rialto Building.

The solution arrived at was to "play a confidence trick" on tower B, making the structure "believe" it is 17 stories taller. Prestressing cables are provided from level 1 to level 38 and stage stressed as tower A construction proceeds. Thereby all columns below level 38 are subject to the same loadings at the same time, and therefore elastic and nonelastic shortening values are relatively consistent for the lifetime of the building (Fig. 4.147).

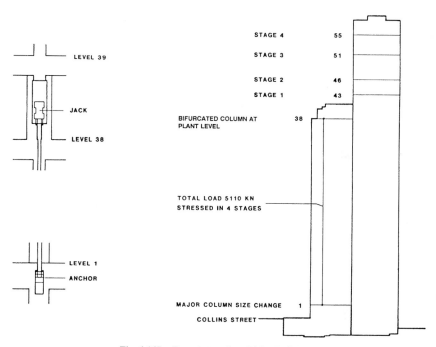

Fig. 4.147   Staged stressing; Rialto Building.

N6E Building
Shinjuku-Ku, Tokyo, Japan

Architect	Nihon Sekkei Inc.
Structural engineer	Nihon Sekkei Inc.
Year of completion	1996
Height from street to roof	189.6 m (622 ft)
Number of stories	46
Number of levels below ground	4
Building use	Offices and retail
Frame material	Steel
Typical floor live load	5 kPa (100 psf)
Basic wind velocity	35 m/sec (78 mph)
Maximum lateral deflection	$H/200$, 100-yr return
Design fundamental period	4.56, 4.75 sec
Design acceleration	35 mg peak, 100-yr return
Design damping	1% serviceability; 2% ultimate
Earthquake loading	$C = 0.0533$
Type of structure	Bundled tube
Foundation conditions	Clay and sand over gravel
Typical floor	
Story height	3.95 m (13 ft)
Beam span	19.6, 16.4 m (64 ft 4 in., 53 ft 10 in.)
Beam depth	800, 600 mm (31.5, 23.5 in.)
Beam spacing	3.2, 3.6 m (10 ft 6 in., 11 ft 10 in.)
Slab	135-mm (5.25-in.) reinforced concrete
Columns	
Size at ground level	600 by 600 mm (24 by 24 in.)
Spacing	3.2, 3.6 m (10 ft 6 in., 11 ft 10 in.)
Core	Framed tube

The plan dimensions of the N6E Building are 92 by 39.2 m (302 by 128 ft), which is quite large (Fig. 4.148). The core location caused eccentricities that could not be reduced using shear walls or bracing systems, so the bundled tube system was adopted to achieve a symmetric structure and to avoid torsional problems (Fig. 4.149). This was done at the expense of reduced span lengths and increased numbers of columns.

The building response was estimated using all available data as well as the along-wind and cross-wind power spectra and cospectra, which vary with the building height. All calculations were done for x, y, and torsional directions.

Fig. 4.148   N6E Building, Tokyo, Japan.

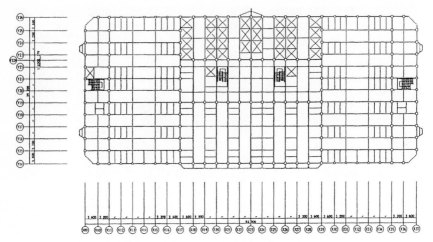

Fig. 4.149   Typical structural plan; N6E Building.

Carnegie Hall Tower
New York, N.Y., USA

Architect	Cesar Pelli and Associates (design)
	Brennan, Beer, Gorman Associates
Structural engineer	Robert Rosenwasser Associates
Year of completion	1989
Height of street to roof	230.7 m (757 ft)
Number of stories	62
Number of levels below ground	1
Building use	Office
Frame material	Concrete
Typical floor live load	2.5 kPa (50 psf)
Basic wind velocity	47 m/sec (105 mph), 100-yr return period
Maximum lateral deflection	Approx $H/500$, 100-yr return
Design fundamental period	4.8 sec E-W; 3 sec N-S; 2 sec torsion
Design acceleration	20 mg peak, 10-yr return period
Design damping	1% serviceability; $2\frac{1}{2}$% ultimate
Earthquake loading	Not applicable
Type of structure	Side-by-side concrete tubes
Foundation conditions	Rock, 4-MPa (40-ton/ft^2) capacity
Footing type	Spread footings
Typical floor	
Story height	3.66 m (12 ft)
Beam span and spacing	Varying
Beam depth	457 mm (18 in.) interior; 762-mm (30-in.) spandrels
Slab	One- and two-way, 230 mm (9 in.) thick
Columns	Size and spacing vary
Material	Concrete, 58 MPa (8400 psi)
Core	Shear walls (part of tubes); thickness varies; concrete as for columns

At 230.7 m (757 ft) in height, Carnegie Hall Tower is the second tallest concrete structure in New York City and the eighth tallest in the world today (Fig. 4.150). With a 15.2-m (50-ft)-wide north face and a 22.9-m (75-ft)-wide south face, which offsets to a 15.2-m (50-ft) face above the forty-second floor, this 62-story structure is the most slender habitable building of this height ever constructed (Fig. 4.151). The structure occupies the narrow site between the five-story Russian Tea Room and the 100-year-old Carnegie Music Hall. The structure's architect, Cesar Pelli Associates, dictated the structural scheme by "sculpting" the structure to complement the existing music hall. The double (side-by-side) tube structural system that resulted was actually defined by filling in all the available spaces between the desired windows with concrete. This resulted in nonuniformity in column size and spacing.

Fig. 4.150   Carnegie Hall Tower, New York.

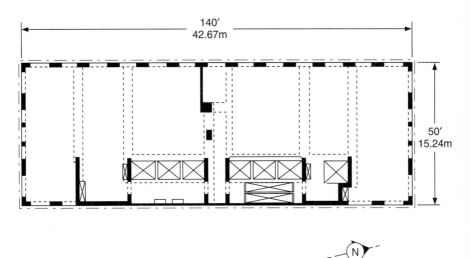

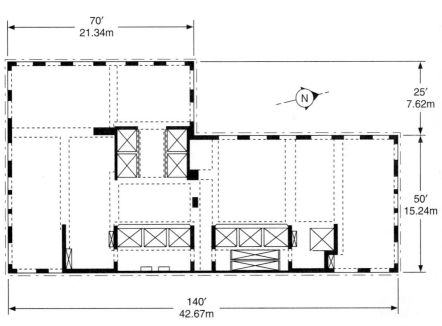

Fig. 4.151   Typical floor plans; Carnegie Hall Tower.

The nonuniformity in the size of the columns at a level was also extended vertically as offsets and larger or smaller window sizes dictated relocation or altered column sizes. Often Vierendeel action was needed to terminate vertical elements at various locations without the benefit of transfer girders. This occurred on the north and south walls and above the forty-second floor for the south half of the west wall, which spans over the enlarged base. Vierendeel action was also required directly above the through-block passage at the ground floor and at several other locations.

A center web (perforated by lobby egress requirements), common to the two side-by-side tubes, was needed to help the north- and south-wall columns to efficiently connect the east flange wall to the west flange wall with minimum shear lag. A Vierendeel column (skipping alternate floors to minimize the lobby obstructions) was introduced to reduce the clear span of the center web. This Vierendeel column is the only interior column in the structure, which otherwise supports all gravity loads by the exterior tube columns and the elevator core walls. The large clear spans of 9.4 m (31 ft) and more between the elevator core and the west wall were spanned with 230-mm (9-in.) slabs and shallow beams 457 mm (18 in.) deep. This framing for gravity loads proved to be more economical than one-way joists or waffle slab construction because it provided more mass to resist uplift forces from wind loads and to reduce building accelerations. It also provided extra height to accommodate mechanical systems so that with a total story height of 3.66 m (12 ft), a ceiling height of 2.7 m (9 ft) was maintained.

The double tube design relies heavily on 760-mm (30-in.)-deep spandrel beams to engage all the vertical supports to resist the wind action and to equalize the stresses due to gravity loads in all supports regardless of their size. The tube's vertical members varied between 480 and 2590 mm (19 and 102 in.) in length (parallel to the exterior) and included a solid concrete wall behind the service core to the east. The structural design considered both the relaxation due to long-term creep and shrinkage of the concrete members and the instantaneous demands of the wind forces.

Enough gravity loads were assembled to eliminate the possibility of tension due to wind in the vertical supports and to let the gravitational loads anchor the structure. A few rock anchors at the west end of the center web were added to enhance the ability of the web to engage the flanges even under larger lateral loads than dictated by the New York City code or the wind tunnel results.

The preliminary design considered both steel and concrete. Control of the perception of motion without auxiliary means such as dampers was found to be attainable only with the concrete alternative because of the larger damping and weight of a concrete structure. However, as a precaution, because of its extreme slenderness, the structure was designed to accommodate a pendulum-type damper. Field measurements, after the structure was topped out, indicated that design predictions were accurate and a damper was not needed. The anticipated accelerations, projected from these load measurements, should not exceed 20 mg for the 10-year return period.

Concrete was pumped in to the full height of the structure. Concrete strength in the columns did not exceed 58 MPa (8400 psi) because the use of silica fume in New York City was still questionable at the time the structure was designed. For this and other slender structures, stiffness, weight, and damping are the important parameters dictating the structure's behavior. The design for acceptable perception of motion often overrides other more mundane design requirements such as strength and stability. This structure together with its earlier slender siblings (Metropolitan Tower, CitySpire, and the Concordia Hotel) are prototypes of the future megastructures of the next generation of tall structures.

**Allied Bank Plaza
Houston, Texas, USA**

Architect	Skidmore Owings and Merrill
Structural engineer	Skidmore Owings and Merrill
Year of completion	1983
Height from street to roof	296 m (972 ft)
Number of stories	71
Number of levels below ground	4
Building use	Office
Frame material	Structural steel
Typical floor live load	2.5 kPa (50 psf)
Basic wind velocity	Unavailable [force = 196 kN/m (13,400 lb/ft) for 100-yr return]
Maximum lateral deflection	$H/500$, 100-yr return
Design fundamental period	Not available
Design acceleration	Not available
Design damping	1% serviceability
Earthquake loading	Not applicable
Type of structure	Perimeter framed tube; diagonally braced core with outrigger trusses
Foundation conditions	Stiff clay
Footing type	Mat 2.9 m (9 ft 6 in.) thick
Typical floor	
Story height	4.0 m (13 ft 1 in.)
Beam span	15.2 m (50 ft)
Beam depth	530 mm (21 in.)
Beam spacing	4.6 m (15 ft)
Material	Steel, grade 250 MPa (36 ksi)
Slab	83-mm (3.25-in.) concrete on 76-mm (3-in.) metal deck
Columns	Built-up, 1016- by 610-mm (40- by 24-in.) perimeter; 610- by 610-mm (24- by 24-in.) interior
Spacing	4.6-m (15-ft) perimeter; 9.15- by 6.1-m (30- by 20-ft) interior
Material	Steel, grade 250 and 350 MPa (36 and 50 ksi)
Core	Braced steel frame, grade 350 MPa (50 ksi)

Allied Bank Plaza was designed to relate strongly to the buildings around it. Situated on a site which is essentially the center of downtown Houston, the building has a major impact on the western facade of the city, which is the most dominant view of its skyline. In form and materials, a design was sought which would be distinctive but would serve to complement and tie together its surroundings. A form that moved and flowed was felt to be appropriate, one that was soft and sheer rather than hard and opaque like the granite and steel rectangular buildings around it (Fig. 4.152).

The resulting semicurved tower was achieved by juxtaposing two quarter-cylinder shafts (Fig. 4.153). The 71-story tower is sheathed in dark green reflective glass, chosen for its sheer quality and responsiveness to light. The combination of plans and curves in the building's design will allow a constant interplay of sunlight on its surface.

Fig. 4.152 Allied Bank Plaza, Houston, Texas. (*Photo by Hedrich-Blessing.*)

Giving the building a human scale was another important aspect of the designer's intentions. Unlike many recent buildings, which are sheathed in reflective glass and appear only as a huge mass, the structure of the Allied Bank Plaza is subtly expressed with vertical and horizontal mullions. A formal portal on the east side of the building provides a sense of entry. Since 65% of the public enters the tunnel-connected downtown buildings at the underground level, Allied Bank Plaza offers the only entrance directly from the street and combines the tunnel with an open-air plaza, including landscaping and a fountain.

A bundled tube frame is the primary lateral system for the 71-story 296-m (972-ft)-tall 186,000-m² (2 million ft²) Allied Bank tower. The shape is formed by two quarter-circles placed antisymmetrically about the middle tubular line. The column spacings are 4.57 m (15 ft) with the usual tree-type construction. The system also uses two vertical trusses in the core, which are connected to the exterior tube by outrigger and belt trusses. Significant improvement in tubular behavior is obtained because of the participation of the trusses. This system, therefore, embodies elements from the framed tube, bundled tube, and truss systems with belt and outrigger trusses. The truss system provides another transverse frame linkage in the curvilinear part to improve its shear lag characteristics.

The structural system for the Allied Bank Plaza tower was selected after study of both steel and composite systems. The system permitted a substantially reduced construction time. The tower's form and slenderness are a radical departure from past rectangular buildings of this height, yet the inherent rigidity of the bundled tube system developed for the tower limited steel weight to 128 kg/m² (26.2 psf).

Design of the frame was based on extensive dynamic modeling of the tower in a wind tunnel study, which confirmed that the wind-shedding form of the tower led to significant reductions in design wind pressure below that experienced by square or rectangular forms. The tower is founded on a 2.9-m (9-ft 6-in.)-thick mat foundation approximately 20 m (65 ft) below grade, which permits utilization of four lower levels for necessary retail, mechanical, and parking functions.

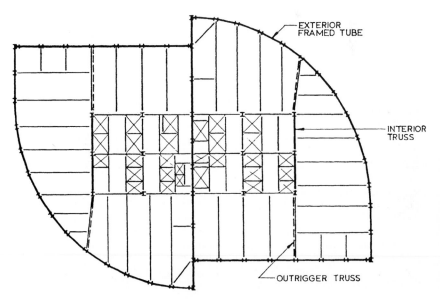

Fig. 4.153   Typical floor framing plan; Allied Bank Plaza.

4.5 HYBRID SYSTEMS

Tall buildings have been traditionally designed to make use of a single type of lateral load resisting system—initially simple moment resisting frames and then shear wall systems and framed tubes. Until the advent of economical, easy-to-use, high-capacity computer hardware and software, structural systems had to be amenable to hand calculation or computer analysis using limited-capacity machines. Nowadays computer capacity is not an issue, and decisions on structural systems are made on the basis of their effects on the appearance and functioning of the building and on its constructability. This is not to suggest that *anything* is acceptable—the engineer must still be aware of the pitfalls of creating abrupt discontinuities in building stiffness, the long-term effects of differential axial shortening, and other side effects of using mixed systems and multiple materials.

An excellent example of a hybrid system is the Overseas Union Bank Center in Singapore. Here a braced steel frame was used because of its lightness, long spanning ability, small member sizes, absence of creep shortening, and, combined with concrete shear walls, for its very cost-efficient contribution to lateral stiffness.

Another type of hybrid system gaining popularity is the concrete-filled steel tube column, where the erectability of a steel frame is maintained, but the cost-effective axial load capacity of high-strength concrete is used. The steel tube provides confinement to the concrete much more efficiently than normal reinforcement does, and it is on the extreme outside, where it is most effective. Of course fire protection must be considered. If the steel tube is considered as sacrificial in a fire, then internal reinforcement sufficient for the reduced loading normally prescribed for the fire limit state must be provided. If external fire protection is provided, then internal reinforcement may not be needed. If concrete can be pumped into the column from the base of each pour, then a number of stories can be concreted at one time and vibration of the concrete is not necessary. Examples of such a system are Casselden Place, Melbourne, and Two Union Square, Seattle.

The trends of modern architecture sometimes force the structural engineer away from convention in a search for a structure that will accommodate aesthetic and functional demands while meeting structural requirements. The result may be a structure which on one face of the building is of a different type than the other faces, as in Georgia Pacific, Atlanta, or a structure with a number of quite different elements forming its lateral load resisting frame, an excellent example being First Bank Place, Minneapolis. Here the engineer has provided a braced steel core connected via outrigger beams to large high-strength concrete perimeter columns, incorporating cast-in steelwork to aid erection and connection. Although this system provides in-plane stiffness, its lack of torsional stiffness required that additional measures be taken, which resulted in one bay of vertical exterior bracing and a number of levels of perimeter Vierendeel "bandages,"—perhaps one of the best examples of the art of structural engineering.

With the advent of high-strength concrete [concrete above 50 MPa or (7000 psi)] has come the era of the "supercolumn," where the stiffness and damping capabilities of large concrete elements are combined with the lightness and constructability of steel frames. High-strength concrete, when it includes silica fume and a high-range water reducer (superplasticizer), exhibits significantly lower creep and shrinkage and is therefore more readily accommodated in a hybrid frame. The relative cheapness of high-strength concrete together with the fact that large members do not require large cranes (or any cranage at all if pumped) means that the columns can be economically designed for stiffness rather than for strength.

The Interfirst Plaza in Dallas (not described in this Monograph) uses supercolumns in conjunction with an almost conventional steel frame, and the Columbia Seafirst Center in Seattle incorporates very large supercolumns connected by steel diagonal members to a braced steel core. Another example, although never built, is the Bank of the Southwest tower in Houston. Here eight giant concrete columns form the chords of four vertical steel megatrusses.

The previous examples suggest that hybrid structures are likely to be the rule rather than the exception for future very tall buildings, whether to create acceptable dynamic characteristics or to accommodate the complex shapes demanded by modern architecture. Hybrid structures are not something to be tackled by the novice engineer armed with a powerful microcomputer and a structural analysis software package, as a sound knowledge and understanding of material behavior (such as ductility, damping, creep, and shrinkage), which is not included in analysis and design packages and mostly not codified, is essential and constructability must be a parallel consideration. However, without hybrid structural systems many of our modern tall buildings may never have been built in their present form.

PROJECT DESCRIPTIONS

Overseas Union Bank Center
Singapore

Architect	Kenzo Tange and Urtec/SAA Partnership
Structural engineer	Meinhardt Asia Pty. Ltd.
Year of completion	1986
Height from street to roof	280 m (919 ft)
Number of stories	63
Number of levels below ground	4
Building use	Commercial, retail, office
Frame material	Steel with concrete walls to stairs and core
Typical floor live load	2.5 kPa (50 psf)
Basic wind velocity	37.7 m/sec (84 mph), 1000-yr return
Maximum lateral deflection	448 mm (17.5 in.)
Design fundamental period	7.3 sec
Design damping	1% serviceability; 3% ultimate
Earthquake loading	Not applicable
Type of structure	Hybrid system of steel frames with concrete walls to increase rigidity
Foundation conditions	Silty sand, sandstone, siltstone, claystone
Footing type	7 caissons 5 to 6 m (17 to 20 ft) in diameter, 100 m (328 ft) deep, belled to 9-m (30-ft) diameter
Typical floor	
Story height	4 m (13 ft 1.5 in.)
Beam span	20.3 m (66 ft 7 in.)
Beam depth	950 mm (37.5 in.)
Beam spacing	4.32 m (14 ft 2 in.)
Material	Steel, grade 50 and 43
Slab	150-mm (6-in.) concrete on metal deck
Columns	
Size at ground floor	800 by 800 mm (31.5 by 31.5 in.)
Spacing	Varies
Material	Steel, grade 55 and 50
Core	Hybrid steel frame with concrete wall zones
Thickness at ground floor	600 mm (24 in.)
Material	Steel, grade 55 and 50; concrete, 45 MPa (6400 psi)

The Overseas Union Bank Center (Fig. 4.154) is a prestige state-of-the-art development designed to house the bank's head office and provide rental office, commercial, and

Fig. 4.154 Overseas Union Bank Center, Singapore.

parking space in Raffles Place, Singapore. The high-rise section is conceived as two visually separate triangle towers (although structurally integral) facing each other on the hypotenuse. A service core and a triangle column in one corner provide support for the higher tower. The lower tower is supported on a smaller triangular column and an L-shaped column. The structure has height-to-width ratios of 10:1 on the south elevation and 8:1 on the north elevation. The high-rise structure provides column-free space throughout its full height above ground (Fig. 4.155).

The high-rise structure is framed using high-yield structural steel. The principal columns are fabricated box columns framing the elevator shafts and flanged T shapes to conform to the wall lines and minimize encroachment into the elevator shaft area.

Simply supported steel trusses 950 mm (37.5 in.) deep spaced at 4.32-m (14-ft) centers in an east-west direction support the large column-free areas. These trusses are designed to act compositely with the concrete floor system.

The floor system consists of a reinforced concrete slab composite with a 63-mm (2.5-in.)-deep ribbed steel deck. The concrete slab is a total of 150 mm (6 in.) thick in order to maintain a sufficient concrete thickness, after reticulation of services, for the required fire separation between levels. Fire protection of the steel frame is provided by lightweight mineral fiber (Figs. 4.156 and 4.157).

The high-rise structure is supported on a total of seven caissons ranging in depth from 96 to 110 m (315 to 360 ft) and in diameter from 5 to 6 m (16.4 to 19.7 ft). The caissons are belled at their base and carry their load in end bearing on solid rock.

Development of the most efficient structural system is the essential prerequisite to optimization of the design. The choice of system dramatically affects the quality of the material required in the design.

The family of structural systems based on the tubular concept has provided the types most widely used to date for high-rise and ultra-high-rise structures. However, it has become necessary to seek new structural systems to respond to changes which have taken place over the last decade, including the very strong influence on high-rise buildings of evolving architectural forms with many large open areas which extend through multiple floor heights.

The decision to use structural steel in lieu of reinforced concrete for the 280-m (919-ft)-high OUB tower was dictated by structural considerations rather than economics (Fig. 4.158). The following are the principal factors that determined the adoption of structural steel in lieu of concrete.

1. The asymmetrical geometry of the structure resulted in higher stresses in the columns supporting the higher triangle than in those supporting the lower triangle. This caused unequal column shortening from creep and a consequent lateral movement of the structure.

2. Differential movement (creep) occurred between the reinforced concrete super-columns in the primary megasystem and the structural steel secondary system within the portal frames of the megastructure.

3. The dimensions of the vertical structural members had become gross, resulting in loss of floor space and presenting substantial planning difficulties, both architecturally and in the distribution of building services.

4. The soil conditions were poor, and a special and costly foundation was necessary. Structural steel keeps the weight down compared to a concrete structure, reducing both the difficulty and the cost of footings.

5. The use of high-yield steel resulted in lighter, smaller, and less costly structural members which would satisfy the system stiffness criteria.

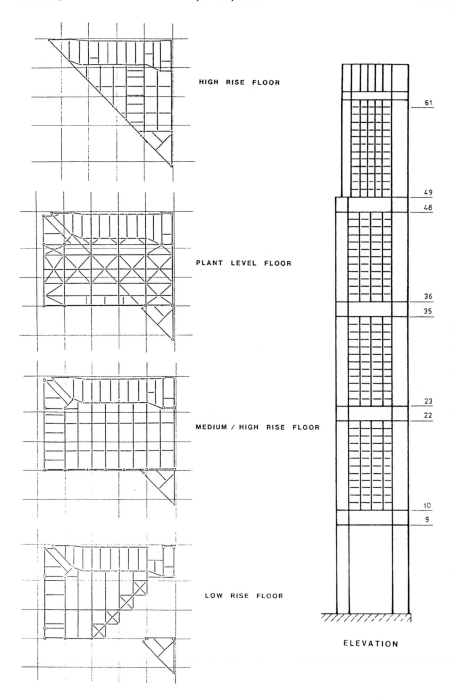

HIGH RISE FLOOR

PLANT LEVEL FLOOR

MEDIUM / HIGH RISE FLOOR

LOW RISE FLOOR

61

49
48

36
35

23
22

10
9

ELEVATION

Fig. 4.155 Framing plans and elevation; Overseas Union Bank Center.

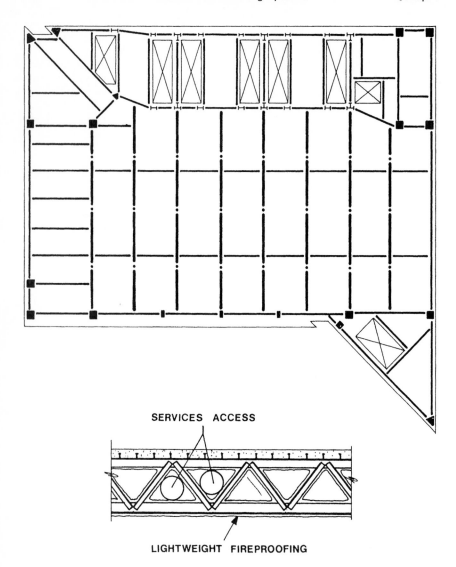

SERVICES ACCESS

LIGHTWEIGHT FIREPROOFING

Fig. 4.156   Floor plan; Overseas Union Bank Center.

The composition of the structure is one where the steel frame provides the skeleton of the structural system, with the bracing and reinforced concrete walled zones acting to increase the rigidity of the building (Fig. 4.159).

The individual elements (steel frame and concrete walls) are both capable of functioning independently in the transfer of vertical loads from the top to the foundations. However, as elements used in conjunction, the concrete provides restraint to the steel, allowing the steel frame to be fully stressed as an isolated component.

Control of differential creep between concrete and structural steel was investigated extensively, taking into consideration axial shortening of the structural steel columns, the construction program, and the bracing of the steel structure during erection. The likely stresses in the concrete elements and the steelwork were considered in both the short term and the long term. The analysis indicated that the optimum was for the concrete elements to follow behind the steelwork by approximately four to five levels. The maximum allowable differential was the concrete elements lagging 24 levels behind the steelwork. The final optimized solution for the OUB structure is a mixed-frame hybrid structure, providing an effective structure utilizing the best properties of steel and concrete to achieve the minimum cost.

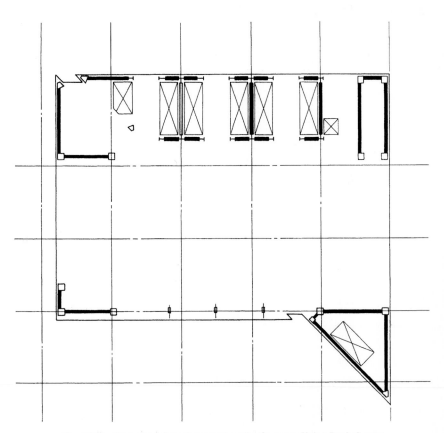

Fig. 4.157   Plan of reinforced concrete walls; Overseas Union Bank Center.

Hybrid structure is worthy of consideration as a conscious design approach. The use of reinforced concrete elements to control the deflection and dynamics of tall steel struc-tures provides an effective alternative structural system that will allow the designer to make full use of the higher allowable stresses of high-yield steels when other bracing systems are inefficient or unacceptable architecturally.

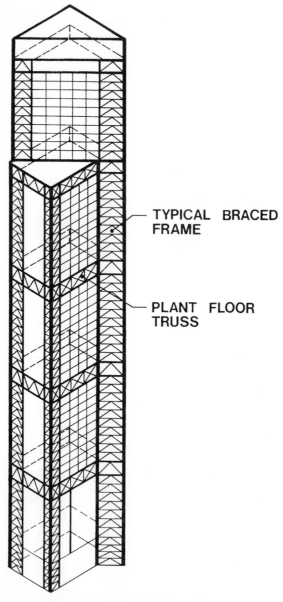

TYPICAL BRACED
FRAME

PLANT FLOOR
TRUSS

Fig. 4.158   Structural steel system; Overseas Union Bank Center.

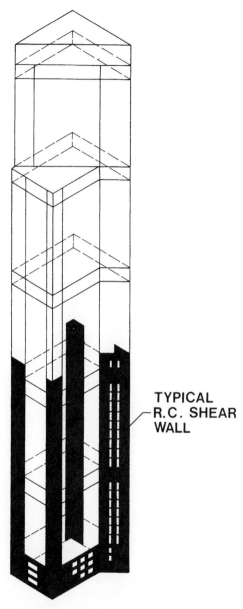

Fig. 4.159   Primary shear wall system; Overseas Union Bank Center.

Citicorp Center
New York, N.Y., USA

Architect	Hugh Stubbins and Associates with Emery Roth and Sons
Structural engineer	LeMessurier Consultants with Office of James Ruderman
Year of completion	1978
Height from street to roof	279 m (915 ft)
Number of stories	60
Number of levels below ground	3
Building use	Office, retail
Frame material	Steel
Typical floor live load	2.5 kPa (50 psf)
Basic wind velocity	41 m/sec (92 mph), 100-yr return
Maximum lateral deflection	$H/600$ at 1-kPa (20-psf) serviceability load
Design fundamental period	6.9, 7.2 sec
Design acceleration	Less than 20 mg peak, 10-yr return
Design damping	Measured 1.1 and 0.9% serviceability increased by TMD to 4% each direction
Earthquake loading	Not applicable
Type of structure	Braced perimeter tube with braced core below 9th floor; TMD at top
Foundation conditions	Manhattan schist
Footing type	Steel base plates on grout on rock
Typical floor	
Story height	3.89 m (12 ft 9 in.)
Beam span	10.3, 12.8 m (33.9, 42.1 ft)
Beam depth	530 mm (21 in.)
Beam spacing	3.81 to 3.91 m (12.5 to 12.84 ft)
Material	Steel, grade 350 MPa (50 ksi)
Slabs	63-mm (2.5-in.) lightweight concrete on 76-mm (3-in.) steel deck
Columns	Pairs of 965 by 762 mm (38 by 30 in.) max by 3421 kg/m (2294 lb/ft) at 5.74-m (18.83-ft) centers at center of each side of building
Material	Steel, grade 350 MPa (50 ksi)
Core	Moment frame above 10th floor, braced frame below
Material	Steel, grade 350 MPa (50 ksi)

This 60-story tower contains an area of 102,200 m^2 (1.1 million ft^2) of the project total of 167,200 m^2 (1.8 million ft^2). The 47.8-m (157-ft) square tower has a dramatic and daring appearance, with all four of its corners jutting out 23 m (76 ft) unsupported from only four exterior columns, one centered on each side, which free-stand for a height of 34.7 m (114 ft) at the base (Fig. 4.160). The central core also supports the tower. This unique structure was not designed this way arbitrarily just to achieve a dramatic effect. The site, a city block in Manhattan, was purchased fully except for St. Peter's Lutheran Church on one corner of the block. The church agreed to sell its air rights, but would

Fig. 4.160   Citicorp Center, New York. (*Courtesy of The Stubbins Association; Photo by Edward Jacoby.*)

allow no columns of the office tower to pass through its facilities, and it required that a new church building be designed and constructed in that corner with its own distinct identity. This last requirement led the architect to place the first office floor more than 46 m (150 ft) above the street.

The most direct and economical way to achieve the 23-m (76-ft) corner cantilevers on each face of the typical tower floor was to provide a steel-framed braced tube with a system of columns and diagonals in compression, channeling the building's gravity loads into a 1.5-m (5-ft)-wide "mast" column in the center of each tower face (Fig. 4.161). The main diagonals repeat in eight-story modules. The compression diagonals are restrained by horizontal tension ties at four-story intervals. This system brings one-

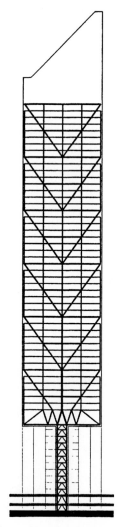

Fig. 4.161 Elevation; Citicorp Center.

half of the tower gravity load down to the four base "legs," one centered on each side. The system, because it repeats on each face of the tower, is also very efficient in resisting wind forces, both shear and overturning, since it forms a complete braced tube. A neat structural touch was the omission of the corner columns at the floor just below the main diagonal intersection with the corner every eight stories. This was to avoid accumulating gravity load in the corner columns and gives unobstructed corner views as a bonus.

An 8.8-m (29-ft)-deep perimeter truss on top of each of the legs carries the gravity loads of the lowest seven floors to the center legs. The wind shear is transferred through the tenth-floor diaphragm at the top chord level of this truss over to the diagonally braced elevator core, which carries it down to the foundation. Wind overturning forces continue from the superstructure mast columns through the legs to the foundation.

The typical office floors are framed with conventional steel beams, with a lightweight concrete slab on electrified underfloor steel deck (Fig. 4.162). The core has moment-connected frames in order to provide a system to deliver floor-by-floor wind forces to the braced tube panel points occurring every fourth story, and to allow shorter unbraced lengths of the main compression members.

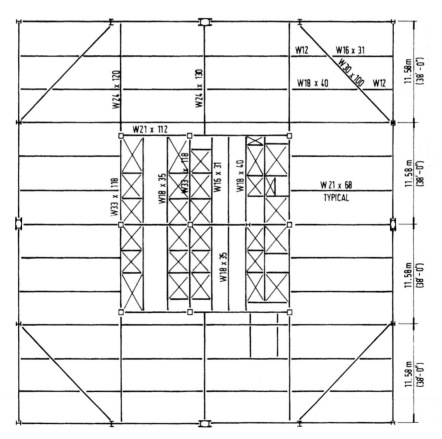

Fig. 4.162   Floor plan; Citicorp Center.

The wind tunnel study for the tower, conducted by the University of Western Ontario, Canada, indicated that persons on upper floors of the tower would experience uncomfortably high lateral sway accelerations in wind storms. In order to reduce accelerations to acceptable levels there were only two possible approaches: add a great deal of mass and lateral stiffness without increasing the natural vibration period, or add to the building's natural damping. The first approach would have cost about $5 million, whereas the second approach would have required increasing the building's damping from about 1 to 4% and designing and constructing the world's first tuned mass damper (TMD) of anywhere near this size. The second approach was adopted at a final cost of less than one-third of the first approach. The initial step was to convince the architect and owner; then the structural engineer had to find a way to actually do it. Fortunately, LeMessurier Associates were able to enlist the technical assistance of Prof. Alan Davenport of the University of Western Ontario, Prof. David Wormley of M.I.T., and the firm of MTS Systems Corporation of Minneapolis. The latter firm provided the detailed mechanical, electrical, and control system design and also constructed the TMD system, with the assistance of HRH Construction of New York, the general contractor. The TMD is located on a dedicated floor at 242 m (793 ft) above grade, near the top of the tower, for maximum effectiveness. The Citicorp tower was designed from the beginning to have the TMD system. The system used includes a moving 373-tonne (410-ton) concrete-mass block that slides biaxially in the north-south and east-west directions on pressurized oil bearings on polished steel plates. The mass is connected to the building structure via long steel boom struts, pressurized nitrogen springs, and hydraulic servo actuators. The lateral stiffness of the spring elements makes the system into a classical passive spring-mass system, which basically is tuned to the same frequency as the tower and acts as a vibration absorber to effectively increase the building's energy absorption, or damping. The TMD reduced accelerations from wind-induced motion by 40 to 50%. It is designed solely to increase occupant comfort. The building is designed for safety and strength as if the TMD were not there. The TMD system has performed very well since its installation and has weathered many wind storms and even a hurricane.

CenTrust Tower
Miami, Florida, USA

Architect	I. M. Pei and Partners
Structural engineer	CBM Engineers, Inc.
Year of completion	1985
Height from street to roof	178 m (585 ft)
Number of stories	48
Number of levels below ground	None
Building use	Office
Frame material	Concrete
Typical floor live load	2.5 kPa (50 psf)
Basic wind velocity	54 m/sec (120 mph)
Maximum lateral deflection	508 mm (20 in.)
Design fundamental period	3.50, 4.50 sec
Design acceleration	Not calculated
Design damping	2% serviceability; 5% ultimate
Earthquake loading	Not applicable
Type of structure	Perimeter partial tube with interior shear walls
Footing type	2.1- to 2.4-m (7- to 8-ft)-thick mat on pre-cast piles
Typical floor	
Story height	3.81 m (12.5 ft)
Beam span	14.6 m (48 ft) max
Beam depth	508 mm (20 in.) with 813-mm (32-in.) haunching
Slab	Concrete joists at 1.8-m (6-ft) centers and 114-mm (4.5-in.) slab
Columns	
Size at ground floor	1600- to 1220-mm (63- to 48-in.) diameter at 4.57-m (15-ft) centers
Material	48-MPa (7000-psi) concrete
Core	Shear wall, 610 mm (24 in.) thick max
Material	48-MPa (7000-psi) concrete

Overlooking Biscayne Bay, the 48-story CenTrust Tower adds a unique shape to the skyline of downtown Miami (Fig. 4.163). The building consists of a 37-story-tall office tower set on top of a block square 11-story parking garage. A quarter-circle in plan, the office tower's arc steps back three times as it rises up. The 90° corner of the quarter-circle is chamfered to create an additional 25.9-m (85-ft)-wide face of the building. The garage also serves Miami's convention center and has a people mover station on its fourth floor. On top of the garage, the building carries a large landscaped area, including a reflection pool.

The building is constructed in reinforced concrete. Floor framing consists of 520-mm (20.5-in.)-deep pan joists, spanning up to 10.7 m (35 ft) and supported on 14.6-m (48-ft)-long haunched girders. Depth of the haunched girders varies from 520 mm (20.5 in.) in the middle to 813 mm (32 in.) at the ends.

The three 4.6-m (15-ft) step backs at the circular face of the building are located at floors 20, 31, and 46, as shown in Fig. 4.164. Conventional girders are used to transfer the columns at floor 46, but at floors 20 and 31 an unusual one-floor-deep bracket is employed to transfer each column. A normal maze of transfer girders would have resulted

Fig. 4.163   CenTrust Tower, Miami, Florida. (*Courtesy of CBM Engineers, Inc.*)

in a loss of lease space at both of these floors. The location of transfer columns and brackets at the twentieth floor is indicated in Fig. 4.165, and a typical one-sided bracket from the perimeter column is shown in Fig. 4.166. The gravity column loads at the twentieth floor range between 13,300 and 17,800 kN (1550 and 2000 tons).

Under column load the bracket requires lateral bracing, which is provided in the form of wall shear panels between floors 19 and 20. Where a wall shear panel aligns with the bracket, compression and tension chord forces are directly anchored in these wall panels. Such tension chords at floors 19 and 30 are prestressed with an effective force of

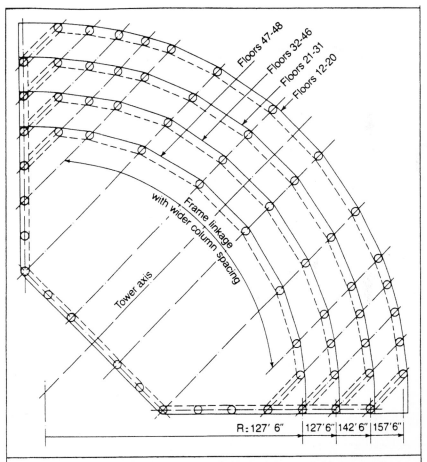

Tower axis is 45 degrees off garage axis. Because the columns of the curved wall describe an arc in the garage, their spacing is wider than those on the straight walls to accommodate the parking bays.

Fig. 4.164   Perimeter column layout; CenTrust Tower.

11,136 kN (1250 tons). For the other brackets, these chord forces are transferred to the wall panels via floor plates acting as in-plane diaphragms.

The floor slab over pan joists is increased from 114-mm (4.5-in.) normal thickness to 190 mm (7.5 in.) at floors 19, 20, 30, and 31 to provide required strength and stiffness for the in-plane diaphragm forces.

A partial framed tube at the perimeter of the tower and minimal shear walls in the core are provided for the lateral load resistance, causing least interruption in the flow of traffic in the garage and a minimum loss of parking spaces. Shear walls are transferred

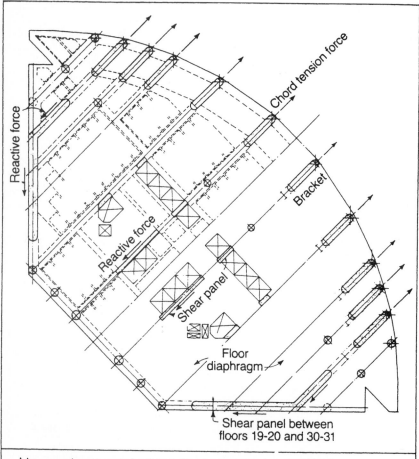

Unusual eccentric transfer brackets at the 19th and 30th floors transfer wind and gravity loads directly to the perimeter columns. Plan of 19th floor is shown here; 30th floor is similar.

Fig. 4.165   Transfer floor plan; CenTrust Tower.

to columns at the tenth floor of the garage to facilitate traffic flow. The partial framed tube is carried through the garage and designed to resist the entire lateral loads in the garage as well. The partial framed tube consists of two channel-shaped frames with columns at 4.6-m (15-ft) centers linked by frames along the circular arc and the chamfered face, with the columns spaced at 8.6 m (28 ft 3 in.).

Columns in the garage are 1067 by 1880 mm (42 by 74 in.) rectangular and 1372 to 1067 mm (54 to 42 in.) in diameter round. Columns in the tower vary from 1067-mm (42-in.) diameter at lower floors to 762-mm (30-in.) diameter at the top. Spandrel beams are 914 mm (36 in.) deep in the tower, but vary in depth at the garage floors from 1372 mm (54 in.) at the three straight sides to 813 mm (32 in.) along the circular arc due to headroom requirements. Concrete strength in columns and spandrel beams ranges from 49 to 28 MPa (7000 to 4000 psi), but is kept at 28 MPa (4000 psi) for the remaining floor framing.

The tower is supported on a 2.1- to 2.44-m (7- to 8-ft)-thick mat foundation bearing on 350-mm (14-in.) square precast piles. Garage columns are founded on spread footings.

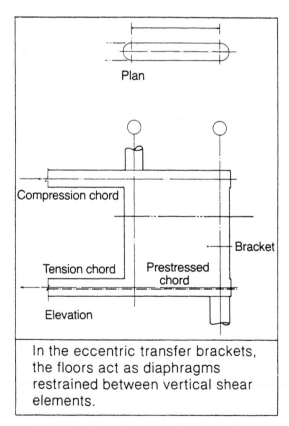

Fig. 4.166   Column transfer detail; CenTrust Tower.

Columbia Seafirst Center
Seattle, Washington, USA

Architect	Chester Lindsey and Associates
Structural engineer	Skilling Ward Magnusson Barkshire, Inc.
Year of completion	1985
Height from street to roof	288 m (947 ft)
Number of stories	76
Number of levels below ground	6
Building use	Retail, commercial, parking, offices
Frame material	Structural steel with composite steel-concrete columns
Typical floor live load	2.5 kPa (50 psf)
Basic wind velocity	34 m/sec (75 mph)
Maximum lateral deflection	483 mm (19 in.), 100-yr return
Design fundamental period	5.3 sec
Design acceleration	20 mg peak, 10-yr return
Design damping	2.5% including dampers for 10-yr return; 2.0% ignoring dampers for 100-yr return
Earthquake loading	$Z = 0.75$, $C = 0.03$, $K = 0.80$
Type of structure	Braced steel core incorporating viscoelastic dampers; triangular core is linked by diagonal steel members at its corners to 3 large steel and high-strength concrete columns
Typical floor	
Story height	3.5 m (11 ft 6 in.)
Slab	50-mm (2-in.) concrete on 50-mm (2-in.) steel deck
Columns	3 major columns, 2.44 by 3.66 m (8 by 12 ft) at ground floor
Material	Concrete, 66 MPa (9500 psi)
Core	Braced-steel rigid frame with arches up to 11 stories tall transferring load to composite columns

This innovative skyscraper has just 73.24 kg/m^2 (14.97 psf) of structural steel and three composite columns of ultra-high-strength concrete. It uses both materials in their most efficient manner. The building is completely framed in structural steel. Wind and earthquake loads are resisted by a structural steel moment resisting braced frame, which is triangular in shape and located in the interior core.

    Exterior windows are unobstructed. Composite structural steel and concrete columns are located at the vertices of the triangular core to carry a large portion of the vertical loads, reduce wind sway, and resist seismic forces. At the base of the structure, these composite concrete columns are 2.44 by 3.66 m (8 by 12 ft) in dimension. The concrete strength is 66 MPa (9500 psi). The sway of the building is limited to $H/600$. The floor framing for the plaza, arcade, and parking levels is a 114-mm (4.5-in.) concrete slab over 76-mm (3-in.) metal deck. Above the plaza level, the floor framing is 50-mm (2-in.) concrete slab over 50-mm (2-in.) deck. All steel floor beams are composite with the concrete slabs.

The skyscraper contains 135,415 m² (1,457,564 ft²) of office space and six below-grade levels of parking for 536 cars with an area of 29,670 m² (319,368 ft²). Public

Fig. 4.167   Columbia Seafirst Center, Seattle, Washington. (*Courtesy of Skilling Ward Magnusson Barkshire, Inc.*)

areas consist of a lobby level containing an area of 1825 m² (19,641 ft²); four arcade shopping levels with a total area of 13,948 m² (150,135 ft²), featuring retail and commercial space; a multilevel shopping arcade which is open 24 hours a day; a multilevel landscaped plaza surrounding the entire office tower; as well as an underground pedestrian tunnel connecting the building to another office building across the street.

Columbia Center's excavation was the deepest ever undertaken in Seattle. It reached 37 m (121 ft) below Fifth Avenue and 21 m (70 ft) below Fourth Avenue. Complicating the task was the requirement to protect an existing five-story office building at the Fourth and Columbia corner of the building. The shoring wall was constructed by drilling 1.2-m (4-ft) holes at 4 m (13 ft) on center to at least 4.3 m (14 ft) below the bottom of the excavation. These holes were filled with lean concrete and a pair of 350-mm (14-in.) wide-flange steel soldier piles. Tiebacks were placed in the normal manner between the pair of vertical soldier piles. 150- and 200-mm (6- and 8-in.) wood lagging was used to support the earth between the pair of soldiers piles spaced 4 m (13 ft) apart. The building structure design underwent the scrutiny of extensive testing in a wind tunnel at the University of Western Ontario, Canada, for both static and aeroelastic loading. The aeroelastic tests measured the twist, sway, base shear, and acceleration of the building. They showed that the building performed very well in the wind, but revealed that the acceleration of the building in a major windstorm might be felt by a portion of the occupants. Viscoelastic dampers to absorb wind energy were added to the moment resisting braced frame to eliminate this possibility of uncomfortable acceleration.

First Bank Place
Minneapolis, Minnesota, USA

Architect	Pei Cobb Freed and Partners, Inc.
Structural engineer	CBM Engineers, Inc.
Year of completion	1992
Height from street to roof	236.5 m (776 ft)
Number of stories	56
Number of levels below ground	3
Building use	Office
Frame material	Steel with concrete supercolumns
Typical floor live load	2.5 kPa (50 psf)
Basic wind velocity	36 m/sec (80 mph)
Maximum lateral deflection	533 mm (21 in.)
Design fundamental period	6.48, 5.26 sec
Design acceleration	24 mg peak, 10-yr return
Design damping	1.25% serviceability; 1.5% ultimate
Earthquake loading	Not applicable
Type of structure	Spine structure, supercolumns, and braced frames with Vierendeel "bandages"
Foundation conditions	Rock, 7.5- to 10-MPa (75- to 100-ton/ft^2) capacity
Footing type	Unreinforced rock footings
Typical floor	
Story height	3.96 m (13 ft)
Beam span	10.97 to 18.28 m (36 to 60 ft)
Beam depth	406 to 838 mm (16 to 33 in.)
Beam spacing	3.05 m (10 ft)
Slab	133-mm (5.25-in.) lightweight concrete on 50-mm (2-in.) metal deck
Columns	2160 mm (84 in.) square at 23.24-m (76-ft) centers
Material	Concrete, 68 MPa (10,000 psi)
Core	Braced spine, A572 steel [350 MPa (50 ksi)]; column size 1067 by 914 mm (42 by 36 in.) to 914 by 610 mm (36 by 24 in.)

A 236-m (776-ft)-tall 56-story chiseled tower is the tallest of three distinct-looking but integral buildings which form First Bank Place (Fig. 4.168). The tower is crowned with a 13.7-m (45-ft)-high circular grid of steel which cantilevers 6 m (20 ft) out from a vertical plane and conceals cooling towers and antennas. At the second floor (the Minneapolis skyway level) the tower connects to buildings on adjacent blocks via two bridges. One of these bridges is a classic tied arch, which is braced from buckling by an inverted pony truss. Adjacent and connected to the tower is the 68-m (224-ft)-tall 14-story atrium building so called because of the six-story 27-m (89-ft)-diameter atrium at

Fig. 4.168   First Bank Place, Minneapolis, Minnesota.

its base. One-fourth of the perimeter of this atrium is a glass wall supported by Vierendeel pipe trusses. Some 12 m (40 ft) above the atrium floor is centered an 18.6-m (61-ft)-diameter ring beam which supports the columns of the lease space floors above the atrium. Filling up the remainder of the L-shaped site is an 18-story 84-m (276-ft)-tall "park" building, which overlooks Hennepin County Government Center Park. Underneath the park building, atrium, and tower is a three-level 450-car basement parking garage. The First Bank Place complex has 130,000 m² (1.4 million ft²) of floor space (Fig. 4.169).

The backbone of the First Bank Place tower is a cruciform-shaped spine anchored by steel and concrete composite supercolumns, which are linked to one another with a vertical shear membrane formed by steel bracing in the core of the building and outrigger beams beyond the core moment-connected into the supercolumns. Characteristic of spine structures, these supercolumns extend uninterrupted the full height of the building. They vary in cross-sectional area along their length from 7 m² (75 ft²) at the base to 4.6 m² (50 ft²) at the top.

Torsional stability for the tower is provided at the perimeter of the building by a dual system of unsymmetrical diagonal bracing and Vierendeel bandages. The single diagonal perimeter braces extend from the third floor to the forty-fifth floor in six-story-high

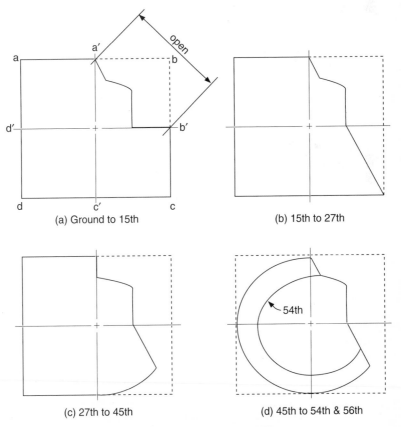

(a) Ground to 15th   (b) 15th to 27th

(c) 27th to 45th   (d) 45th to 54th & 56th

Fig. 4.169   Floor plan; First Bank Place.

sections. Spandrel beams moment-connected through these diagonals, along with the supercolumns, restrain the tendency of these unsymmetrical bracings to deflect horizontally under gravity loads. The three-story-deep Vierendeel girder bandages, which are provided at floors 12–15, 24–27, and 42–45, restrain the warping, which would otherwise occur in the open section composed of the cruciform spine and perimeter braces. These bandages triple the tower's torsional stiffness and increase its lateral stiffness by 36%. In addition, the bandages are used to transfer gravity loads to supercolumns and corner columns, thus increasing the efficiency of the building's overturning resistance.

Other Vierendeels are used in the building to eliminate transfer girders and increase the building's lateral stiffness. A 12-story Vierendeel spans along an exterior face of the building between a supercolumn and a corner column, transferring a column which supports 28 floors of load. Above the forty-fifth floor of the tower there is a nine-story-tall circular Vierendeel girder which frames into supercolumns. The curved Vierendeel not only increases the lateral and torsional stiffness of the top of the building, but also allows the circular-shaped portion of the building to sit atop the square shape below without extending additional columns down through the lease space.

The structural system was chosen over a perimeter braced frame or a moment frame to achieve a column-free exterior facade for the building. The presence of composite concrete columns enhanced the overturning resistance of the building and achieved overall economy for the structure.

A572 grade 50 steel was used for columns and beams that were controlled by strength criteria, and A36 steel was used for members controlled by stiffness criteria. The supercolumns utilized 69- and 55-MPa (10,000- and 8000-psi) concrete. The steel column base plates bear on the top of the concrete basement garage columns, which support the posttensioned flat-plate garage floors. Special analysis was performed to ascertain the effects of restraint on the posttensioned slabs due to the presence of large concrete columns supporting the tower loads and perimeter basement walls. All building columns sit on individual footings which bear on rock.

Three-story-tall Vierendeel bandages were provided along line CC' and also along $B'D'$ (Fig. 4.170). The strategically placed bandages not only provided essentially column-free exterior spans along face CC', but also improved the torsional resistance of the building dramatically, with optimum use of the structural steel. The perimeter circular Vierendeel above the forty-fifth floor provided both lateral and torsional resistance to the entire frame.

The lower basement floors were designed as posttensioned concrete flat-plate floors. The posttensioned construction was essential to control cracking in floor slabs because of the cold, snowy winters of Minneapolis.

The building was analyzed in a three-dimensional finite-element computer model for the following loading conditions:

1. Sequentially applied dead load consistent with the construction sequence of the building

2. Live load

3. Three-directional (x, z, and θ) wind loads dynamically determined from wind tunnel study with appropriate combinations

4. Creep and shrinkage of concrete columns

5. Temperature gradients and differential temperature on concrete columns

During the design, the members were checked for 99 load combinations.

In the construction sequence, the composite concrete column was allowed to lag 12 floors behind erected structural steel and six floors behind the concreted slabs on metal

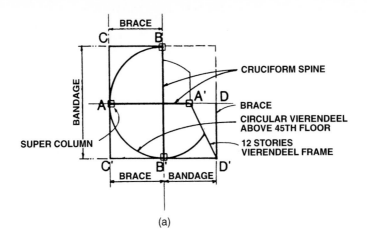

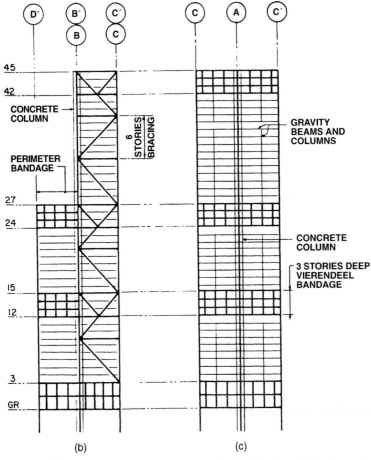

Fig. 4.170 First Bank Place. (*a*) Structural system. (*b*) External bracing. (*c*) Warping-restraining perimeter bandages.

deck. Because of the presence of unsymmetrical exterior bracing, localized bandages, and the free-spanning Vierendeel above floor 45, the structure was analyzed to establish its performance during the erection process. Both lateral and vertical displacements along with strength were checked.

The strategically placed perimeter warping-restraining bandages improved the torsional performance of the structure dramatically. This is evidenced by the comparison of the torsional rotation (Fig. 4.171) and the lateral displacement (Fig. 4.172) of the structure due to wind in the x direction, with and without the bandages.

The presence of the three-story-deep perimeter bandages created a localized horizontal shift in the center of rigidity of the lateral resistance of the structure and thereby

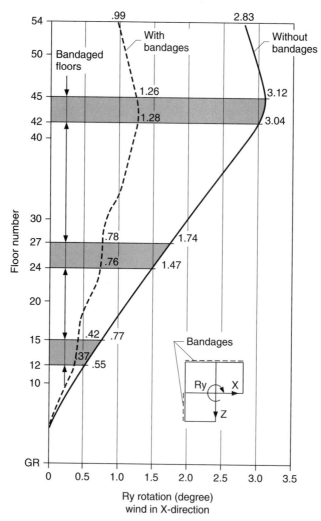

Fig. 4.171   Torsional rotation; First Bank Place.

produced interstory in-plane diaphragm stresses. The associated floor diaphragms were analyzed for in-plane shear and reinforced accordingly.

The building was also analyzed for the reduction in column and diaphragm stiffnesses due to cracking of the concrete and the uncertainty of the effective modulus of elasticity.

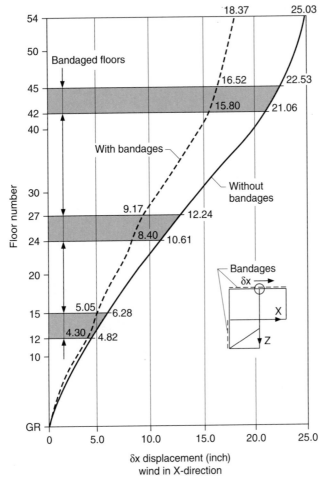

Fig. 4.172   Lateral displacement; First Bank Place.

Two Union Square
Seattle, Washington, USA

Architect	NBBJ
Structural engineer	Skilling Ward Magnusson Barkshire, Inc.
Year of completion	1990
Height from street to roof	220 m (720 ft)
Number of stories	56
Number of levels below ground	4
Building use	Office, retail
Frame material	Steel with composite columns
Typical floor live load	2.5 kPa (50 psf)
Basic wind velocity	34 m/sec (75 mph)
Maximum lateral deflection	312 mm (12.3 in.), 100-yr return
Design fundamental period	6 sec
Design acceleration	20 mg peak, 10-yr return
Design damping	2.1%/10-yr return including damping devices; 2.0%/100-yr return ignoring damping devices
Earthquake loading	$Z = 0.75$, $C = 0.03$, $K = 0.8$

Two Union Square building provided the construction industry with many new concepts, materials, and techniques (Fig. 4.173). By locating the earthquake and wind resisting elements in the interior core walls, the architect had freedom that contributed to its design. Two Union Square represents a union of business and community. It comprises a 56-story office tower with over 93,000 m^2 (1 million ft^2) of office space, four levels of underground parking for 1100 cars, and a remarkable 4,000-m^2 (1-acre) three-level plaza with large open spaces, retail shops, and restaurants.

The design team was faced with a number of unique challenges by this complex curvilinear-shaped building, including meeting the desires of the owners for a signature building—one that would reflect the flavor of the Northwest and provide inviting public spaces at the base of the tower. Particularly challenging was the constraint-filled site, which included existing structures on two sides, an active interstate freeway adjacent, a city street over the base and under the tower, and requirements for a complex water feature with creek, waterfall, and large boulders stair stepping through the plaza.

Among the many technical accomplishments that increased performance, shortened construction time, and reduced structural costs from $28 to $18 million are the most advanced application of a composite system, the first to utilize steel pipes filled with a world-record-breaking high-strength 131-MPa (19,000-psi) concrete, the most efficient viscoelastic dampers to control building movement, and unequaled exterior column spacings of up to 14 m (46 ft), providing sweeping views of the city and Puget Sound.

A complex curvilinear structure, requiring special design considerations for wind engineering, floor framing, and foundations, including a 21-m (70-ft)-deep excavation, and sited in a seismically active area (seismic zone 3), the design provided new technology and unmatched cost efficiency. But more than just an innovative structure, this landmark building provides a warm, friendly atmosphere with retail stores, restaurants, needed parking for downtown shoppers, a respite in a busy downtown area, scenic viewpoints, an extension of an important urban park, and a new network of pathways for the adjoining neighborhoods. Its exceptional design has won widespread architectural praise and public popularity and received the Grand Award for Engineering Excellence from the American Consulting Engineers Council in 1990.

Fig. 4.173   Two Union Square, Seattle, Washington. (*Courtesy of Skilling Ward Magnusson Barkshire, Inc.*)

First Interstate World Center
Los Angeles, California, USA

Architect	I. M. Pei and Partners
Structural engineer	CBM Engineers, Inc.
Year of completion	1990
Height from street to roof	310.3 m (1018 ft)
Number of stories	75
Number of levels below ground	2
Building use	Office
Frame material	Structural steel
Typical floor live load	2.5 kPa (50 psf)
Basic wind velocity	31 m/sec (70 mph)
Maximum lateral deflection	584 mm (23 in.), 100-yr return
Design fundamental period	7.46, 6.91 sec
Design acceleration	23 mg peak, 10-yr return
Design damping	1.25% serviceability; 1.5% ultimate
Earthquake loading	$C = 0.03$, $K = 0.8$
Type of structure	Perimeter ductile tube with chevron braced core
Foundation conditions	Shale
Footing type	Spread footings
Typical floor	
Story height	4.04 m (13 ft 3 in.)
Beam span	16.76 m (55 ft)
Beam depth	610 mm (24 in.)
Beam spacing	4 m (13 ft)
Slab	133- or 159-mm (5.25- or 6.25-in.) lightweight concrete on metal deck
Columns	1067- by 610-mm (42- by 24-in.) WF section, grade 350 MPa (50 ksi)
Spacing	6.1 to 7.6 m (20 to 25 ft)
Core	Braced steel; column size at ground floor 1230 mm (48 in.) square, 6308 kg/m (4230 lb/ft)

Called a signature building for the city of Los Angeles, the granite clad, 75-story building with its serrated facade (Fig. 4.174) rises 310.3 m (1018 ft) above street level. It contains about 130,000 m² (1.4 million ft²) of office space. At present, it is the tallest building in seismic zone 4 or its equivalent in the world.

The base of the tower is embellished by Spanish steps with water runnels, fountains and landscaped areas. These steps are seismically isolated from the tower structure and bridge the elevational difference of approximately 15 m (50 ft) in the surrounding area along the north to south axis of the tower.

The structural system for the tower is an all steel dual system comprising an interactive braced core and a perimeter ductile moment frame. The braced core, anchored at its corners by steel box columns, is 22.5 m (73.8 ft) square. The box columns weighing a maximum of 6308 kg/m (4320 lb/ft) at the base carry a maximum design gravity load of 100,000 kN (11,000 tons). Two-story chevron braces free span each of four sides of the core. In order to achieve an efficient lateral load resisting structural frame, floors free span up to 16.76 m (55 ft), loading the interior core and the perimeter frame columns. These columns perform the dual function of supporting gravity loads and participating in resisting axial loads due to overturning moments generated by lateral wind and seismic loads. Ductility is provided in the structure by the welded perimeter ductile moment resisting frame.

The structure is designed to remain essentially elastic for an anticipated maximum credible earthquake of magnitude 8.3 on the Richter scale at the nearby San Andreas

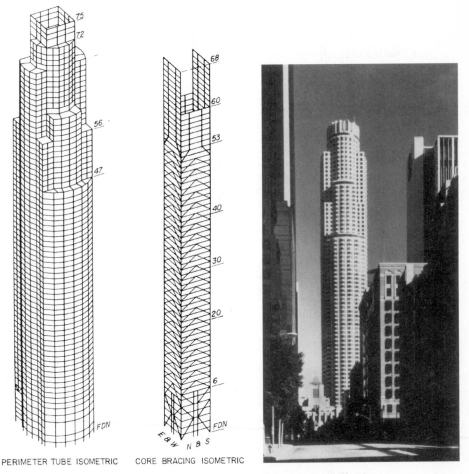

PERIMETER TUBE ISOMETRIC   CORE BRACING ISOMETRIC

Fig. 4.174   First Interstate World Center, Los Angeles, California.

fault. Apart from analyzing the structure for a conventional 5 percent damped response spectrum for the maximum credible earthquake, the following special analysis and design features were introduced.

1. Since two-story chevron bracing was used for the first time in the seismic region, redundancy in the gravity structural load path was examined for an accidental buckling of a diagonal.

2. The structural members, both beams and columns, were not only designed for the ground motion along the two orthogonal principal axes of the structure, but also were checked for the directional maxima due to omnidirectional seismic motion.

3. Time history analysis was conducted primarily to determine maximum interstory drift and the absolute maxima for the horizontal acceleration at floors. The maximum interstory drift was used in the design of the curtain wall, whereas the acceleration data was used for the design of floor-mounted equipment such as elevator machines and water tanks. Time history analysis was also conducted for vertical acceleration. Besides creating overturning effects at the transfer floors, an amplification of vertical acceleration could lead to a plunging failure in the transfer girders. The analysis was essential to preclude such a failure mode.

4. In order to establish a load deflection curve and global ductility limits, a monotonically increasing symmetric nonlinear lateral load analysis was conducted.

5. The criterion for wind motion was set at around 23 mg for peak horizontal acceleration during a once in 10 years wind storm. The lateral modes of vibration were adjusted in a way, not only to achieve the occupant comfort at the top occupied floor for the 10 year wind storm, but which would not increase the lateral response to seismic motion.

Sixteen critical joints in the braced frame were mechanically stress relieved by using the Leonard Thompson vibrating method of stress relief. Special welding and testing procedures were established for all welded connections.

The structure is founded on shale rock with an allowable load bearing capacity of 720 kPa (7.5 tons/ft^2). The core of the structure is supported on a 3.5 m (11.5 ft) thick concrete mat, and a perimeter ring footing is used for the ductile frame.

Hongkong Bank Headquarters
Hong Kong

Architect	Foster Associates
Structural engineer	Ove Arup and Partners
Year of completion	1985
Height from street to roof	180 m (590 ft)
Number of stories	45
Number of levels below ground	4
Building use	Office, banking
Frame material	Structural steel frame; composite steel and concrete floors
Typical floor live load	5 kPa (104 psf) with some local increases
Basic wind velocity	64 m/sec (144 mph), 50-yr return, 3-sec gust
Design fundamental period	4.4 sec
Design acceleration	20 mg peak, 10-yr return period (typhoon event)
Design damping	1% serviceability
Earthquake loading	Not applicable
Type of structure	Steel mast joined by suspension trusses acting in portal frame action
Foundation conditions	Loose fill over marine deposits and decomposed granite bedrock; granite bedrock up to 40 m (131 ft) below ground
Footing type	Machine- and hand-dug caissons to rock
Typical floor	
Story height	3.9 m (12.8 ft)
Beam span	11.1 m (36 ft)
Beam depth	900, 406 mm (35.5, 16 in.) steel
Slab	100-mm (4 in.) reinforced concrete
Columns	
Size at ground floor	1.2-m (4-ft)-diameter in groups of four
Spacing	8 groups in total on grid of 38.4 by 16.2 m (126 by 53 ft)
Material	Steel, grade 50
Core	None

The 20-m (65-ft)-deep basement of the Hongkong Bank (Fig. 4.175) was constructed using a perimeter diaphragm wall and top-down construction techniques. The superstructure is constructed using structural steel and composite steel floors. Stability is provided by masts, linked at five levels by trusses, the complete system acting as a five-level unbraced sway frame. Each mast comprises four tubular steel columns linked by horizontal box-section beams to create a Vierendeel system (Figs. 4.176 and 4.177).

Fig. 4.175   Hongkong Bank Headquarters, Hong Kong. (*Courtesy of Ove Arup and Partners.*)

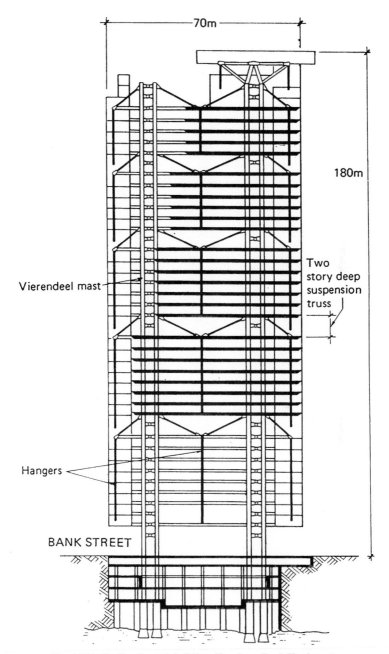

Fig. 4.176   Section through building; Hongkong Bank Headquarters.

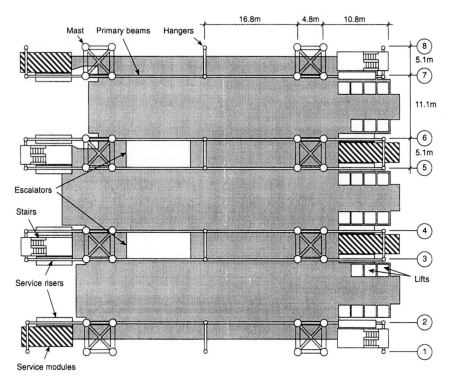

Fig. 4.177   Typical floor plan, Hongkong Bank Headquarters.

4.6 CONDENSED REFERENCES/BIBLIOGRAPHY

AISC 1983, *Modern Steel Construction*

AISC 1987, *One Liberty Place—Efficiency and Elegance in the Cradle of History*

Architecture 1988, *Exploring Composite Structures*

Architecture 1988, *Two Union Square*

Architecture 1990, *High Strength*

Architecture and Urbanism 1991, *Two Union Square*

ASCE 1986, *Computer Cuts Tower Steel*

ASCE 1990, *Aussie Steel*

Australia Post Publ. 1988, *Chifley Square on the Move Structures*

Building 1990, *Double Strength*

Building Design and Construction 1984, *Building Design and Construction*

Civil Engineer 1987, *Concrete Strength Record Jumps 36%*

Concrete Today 1989, *Always Something New in Concrete*

Construction Specifier 1988, *Innovative Composite Construction*

Construction Steel 1990, *The Many Faces of the Bond Building*

Drew 1990, *Rialto Towers Project Seismic Response Analysis and Evaluation*

Engineering News Record 1988, *Sydney Skyscraper Sets Sail*

Engineering News Record 1989, *19,000 psi*

Engineering News Record 1990, *Innovative Techniques*

Engineering News Record 1991, *Sydney Tower Tests Australians*

George 1990, *Wellington's Winds Shaped the Capital's Tallest Building*

Gillespie 1990, *Design and Construction of Steel Framed High-Rise Buildings*

Grossman 1985, *780 Third Avenue, The First High-Rise Diagonally Braced Concrete Structure*

Grossman 1986, *Behavior, Analysis and Construction of a Braced-Tube Concrete Structure*

Grossman 1989, *Slender Structures—The New Edge (II)*

Grossman 1990, *Slender Concrete Structures—The New Edge*

Horvilleur 1992, *Design of the Nations Bank Corporate Center*

Hose 1990, *Structural Design for the Rialto Towers*

Itoh 1991, *Wind Resistant Design of a Tall Building with an Ellipsoidal Cross Section*

Journal of Wind Engineering and Industrial Aerodynamics 1990, *Optimisation of Tall Buildings for Wind Loading*

Kurzeme 1985, *Deep Caisson Foundations for OUB Centre, Singapore*

Kurzeme 1990, *The OUB Centre Tower Foundations, Singapore*

Meinhardt 1984, *Superstructure Design for the Overseas Union Bank Building, Singapore*

Meinhardt 1990, *The OUB Centre—Quality Delivery*

Melbourne 1985, *Aeroelastic Model Tests and Their Application for the OUB Centre, Singapore*

Platten 1986, *Postmodern Engineering*

Platten 1988, *Momentum Place: Steel Solves Complex Geometries*

Taranth 1988, *Structural Analysis and Design of Tall Buildings*

5

Special Topics

5.1 DESIGNING TO REDUCE PERCEPTIBLE WIND-INDUCED MOTIONS

The structural systems for tall buildings are more often controlled by the need to restrict response to wind action at serviceability levels than the need to provide resistance at ultimate limit-state conditions. This section will deal specifically with the criteria related to human occupancy comfort and the design procedures used to establish the response of a building to wind action and the sensitivity to aerodynamic shape, damping, stiffness, mass, and mode shape. Some mention will be made of implications to ultimate limit-state design as something which tends to be dealt with after the system has been designed to cope with the serviceability requirements.

1 Response and Excitation Mechanisms

The response of tall buildings to wind action can be conveniently separated into along-wind and cross-wind motion in relation to the two distinctly separate excitation mechanisms. The total response is, of course, a response to both these motions superimposed on each other, which results in a random, and sometimes roughly elliptic, motion of the top of the building.

The along-wind response is made up of a mean component and a fluctuating component. The addition of these two components is important to the determination of ultimate limit-state loads, but it is only the fluctuating component which gives rise to accelerations that affect occupancy comfort. For the cross-wind response, the mean component is usually very small, with the fluctuating component dominating the response. The fluctuating component of the along-wind response is primarily driven by fluctuating pressures on the upstream face, which are caused by the fluctuating wind speeds in the incident turbulent flow. These pressure fluctuations are converted to along-wind response of the building through a combination of quasi-steady response to low-frequency components and narrow-band resonant response, primarily in the first mode. The cross-wind fluctuating response is primarily a narrow-band resonant response to the fluctuating pressures on the streamwise surfaces caused by the fluctuating vortices shed into the wake. It is referred to as wake excitation where buildings are concerned in order to distinguish it from the narrow-band vortex excitation of slender structures such as chimneys. The mechanisms really are the same, but the broad-band nature of the cross-wind pressure fluctuations normally associated with buildings is due to both the effects of turbulence and the intermittent reattachment of the separated shear layers onto the streamwise faces of the building. Typical along-wind and cross-wind response traces and spectra are given in Fig. 5.1, which illustrates the response characteristics described.

Later in this section analytical methods will be given to permit prediction of the along-wind and cross-wind responses. However, to permit some further description of the fluctuating components that are important to serviceability and ultimate limit-state considerations, it is helpful to refer to a diagrammatic representation of the along-wind and cross-wind forcing spectra, as is presented in Fig. 5.2.

Tall buildings typically have serviceability and ultimate limit-state operating values of reduced velocity in the range of 2 to 10. For example, a 300-m (984-ft)-high building with a width b of 50 m (164 ft) and first-mode frequency n of 0.15 Hz might have serviceability and ultimate limit-state design mean wind speeds at the top of the building

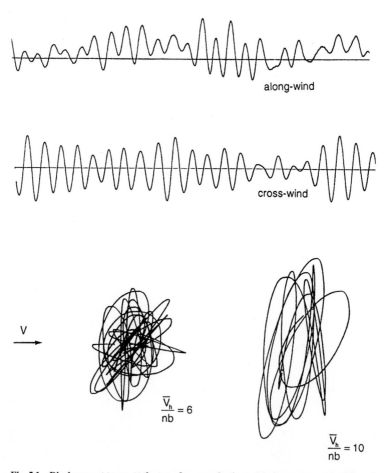

Fig. 5.1   Displacement traces at the top of an aeroelastic model of a square tower: $h/b = 7$.

height of 26 and 45 m/sec (85.3 and 147.6 ft/sec), respectively, which gives, for serviceability,

$$\frac{\overline{V}}{nb} = 3.5 \tag{5.1}$$

and for the ultimate limit state,

$$\frac{\overline{V}}{nb} = 6.0 \tag{5.2}$$

With reference to Fig. 5.2 it can be seen that buildings operating in the low reduced-velocity range are not likely to have occupancy comfort problems. At higher operating

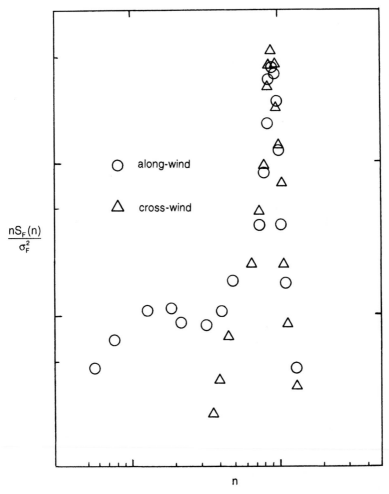

Fig. 5.2   Along-wind and cross-wind force spectra for model square tower; $h/b = 7$; $\overline{V}_h/nb = 10$.

reduced velocities the normalized-force spectral value available to excite the cross-wind motion rises rapidly to a peak while that for along-wind motion is increasing only slowly. This highlights the fact that buildings which do have difficulties in meeting occupancy comfort acceleration criteria inevitably are operating at high reduced velocities, and the dominant cause of the problem is cross-wind motion, as along-wind accelerations are only likely to exceed cross-wind accelerations at reduced velocities below about 2.

2   Acceleration Criteria for Occupancy Comfort

A summary of the development of acceleration criteria has been given in Melbourne and Palmer (1992), and some will be repeated here to provide a background to the recommended acceleration criteria.

Acceleration criteria to achieve occupancy comfort in tall buildings have received somewhat varied attention over the past 20 years. The pioneering work of Chen and Robertson (1973) gave valuable information about human perception of sinusoidal excitation as a function of frequency, and they suggested using an occupancy sensitivity quotient which defines the ratio of the tolerable amplitudes of motions to the threshold motion for half the population. The work by Reed (1971) gave the first full-scale evaluation of occupants' responses to accelerations on two buildings and the first criteria in terms of standard deviation of acceleration for a return period for the frequencies of those buildings. Irwin, in a series of papers, further studied the responses of humans to sinusoidal acceleration over a range of frequencies. Through a number of unpublished full-scale studies of crane operators and building occupants he was primarily responsible for the standard deviation acceleration criteria in ISO 6897. He focused these on tall buildings in Irwin (1986). In North America some use appears to be made of an unreferenced peak acceleration criterion of 20 mg once in 10 years, with no reference to frequency.

Melbourne (1980), using the earlier work of Chen and Robertson (1973), Reed (1971), and some full-scale observations, developed criteria for tall buildings undergoing normally distributed oscillations, as distinct from the sinusoidal oscillations used in the human perception experiments. This development related the accelerations of the normally distributed and sinusoidal motions on the basis that it was the peak accelerations that were the most important. Chen and Robertson had suggested that even the third derivative of displacement (jerk, as they termed it) could be significant. The criterion from Melbourne's work was that structures should be designed such that peak horizontal accelerations do not exceed 10 mg per year on average for frequencies in the range of 0.2 to 0.3 Hz, with the implication that motions perceptible by most people would occur during one or two storms per annum. Later Melbourne and Cheung (1988) commenced the development of frequency-dependent criteria for peak accelerations, to be used for buildings undergoing complex motions and for different distributions.

The development of acceleration criteria referenced to peak accelerations is based on the conclusions of Chen and Robertson (1973) that it is the second or even third derivative of displacement which is most relevant to human comfort, or rather discomfort. Criteria based on the standard deviation of acceleration ignore the probability distribution of the peak accelerations, which varies greatly between a sine wave and a normally distributed process, and very significantly, for example, between the cross-wind response of a building operating near the peak of the cross-wind force spectrum and the along-wind response—hence the need to base acceleration criteria for occupancy comfort in tall buildings on peak acceleration rather than the standard deviation of acceleration.

Horizontal acceleration criteria in terms of the standard deviation of acceleration for the worst 10 consecutive minutes in a 5-year return period for buildings as a function of frequency have been given by Irwin's E2 curve and by curve 1 in Fig. 1 of ISO 6897. This curve can be fitted by the equation

$$\sigma_{\ddot{x}} = \exp(-3.65 - 0.41 \ln n) \tag{5.3}$$

where $\sigma_{\ddot{x}}$ = standard deviation of acceleration in horizontal plane
n = frequency of oscillation with approximately normal distribution

This curve is not so much described by Irwin (1986) as a criterion per se, but ISO 6897 describes it as a satisfactory magnitude and the level at which about 2% of the occupants will comment adversely. Irwin notes that in general the criterion for satisfactory magnitudes of vibration in each of a number of categories (which he describes) is based on a minimum adverse comment level of 2% of the population involved. This definition of acceptability was also suggested earlier by Reed. It has been shown (Melbourne, 1988) that Reed's standard deviation criterion of 5 mg for a 6-year return period, Melbourne's 1-year return-period peak acceleration of 10 mg, and the North American 10-year return-period peak acceleration of 20 mg in the frequency range of 0.2 to 0.3 Hz are all in reasonable agreement with Eq. 5.3. Hence it was concluded that with this degree of agreement it was reasonable to use Eq. 5.3 as a base for developing frequency-dependent peak acceleration criteria for horizontal motion in buildings for occupancy comfort.

The peak acceleration has been obtained from the standard deviation expression, Eq. 5.3, on the assumption that it relates to a normally distributed process, using

$$\hat{\ddot{x}} = g\sigma_{\ddot{x}} \tag{5.4}$$

where g = peak factor, $\approx \sqrt{2 \ln nT}$ for a normally distributed process (Melbourne, 1977)
n = frequency
T = duration, sec; that is, for 10 min, $T = 600$ sec

This gives, for a 5-year return period,

$$\hat{\ddot{x}} = \sqrt{2 \ln nT} \exp(-3.65 - 0.41 \ln n) \tag{5.5}$$

It must be noted that to evaluate a peak acceleration from the earlier data of Chen and Robertson and of Irwin, based on experiments with sinusoidal motion, $g = \sqrt{2}$. To generalize this acceleration criterion for other return periods, it is necessary to choose relationships for wind speed versus return period and response versus wind speed.

For the former, a Gumbel line fit to daily data typical of those for areas in which extremes are generated by both thunderstorm and extensive pressure systems has been used as follows:

$$\hat{V} = 30 + 3 \ln R \tag{5.6}$$

where R is the return period in years. For the latter, a response based on $\hat{V}^{2.5}$ has been used as a compromise between along-wind and cross-wind response. A fit to the factor generated by these assumptions is as follows:

$$\frac{\text{Response for return period of } R \text{ years}}{\text{Response for return period of 5 years}} = 0.68 + \frac{\ln R}{5} \tag{5.7}$$

Combining these gives the peak acceleration criterion for occupancy comfort in buildings as

$$\hat{\ddot{x}} = \sqrt{2 \ln nT}\left(0.68 + \frac{\ln R}{5}\right)\exp(-3.65 - 0.41 \ln n) \tag{5.8}$$

Plots of the acceleration criterion are given as a function of frequency in Fig. 5.3 for a period of 10 min of maximum wind in a return period of R years. The period of 10 min has been used both to fit in with the original curves of Irwin and of ISO 6897, and because it is typical of a period of maximum response in areas dominated by thunderstorm activity and where mean design wind speeds tend to be worked backward artificially from peak wind-speed data. For regions where the maximum response may occur through longer periods, such as 1 hour, the maximum hourly mean wind speed will be less than the maximum 10-min mean wind speed, and the value of T in Eqs. 5.5 and 5.8 would increase to 3600 sec.

3 Determination of Response

At the design stage estimates of the response of a building are required to determine serviceability acceleration levels, equivalent static ultimate limit-state base moments, and moment and shear force distributions. These estimates may be obtained analytically, from wind tunnel measurements, or from a combination of the two. The wind tunnel derivation of these design data will be given elsewhere in this Monograph series. For this

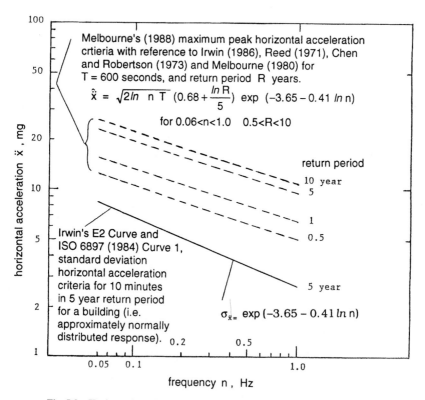

Fig. 5.3 Horizontal acceleration criteria for occupancy comfort in buildings.

section, a description of the analytical approaches will be given, albeit heavily empiri-
cally supported in places.

Along-Wind Response.   The gust factor approach pioneered by Davenport (1967) and
Vickery (1966, 1969) provides the simplest means of estimating the along-wind response of a building and the equivalent static load to produce the peak response. Ver-
sions of this approach have been developed in a number of the world's wind-loading
codes. In particular, the Australian code AS1170.2-1989 has a version in which all the
parameters are given in equation form.

As the gust factor approach is in such general use, there is no need to develop it here,
particularly as it relates to the determination of ultimate limit-state design data. However, so that comparisons may be made between along-wind and cross-wind serviceability accelerations, it will be of help to develop the along-wind equations here. The
evaluation of the along-wind response is divided into background and resonant response
components. The background, or quasi-steady, response is at random and relatively low
frequencies. It is the narrow-band resonant response component, which generates the
majority of the along-wind acceleration at the top of a building. Using the gust factor
approach, the peak acceleration at the top of a building for resonance in a fundamental
bending mode may be obtained from

$$\hat{\ddot{x}} = \hat{x}_{res}(2\pi n_0)^2$$

$$= G_{res}\frac{M}{M_I}(2\pi n_0)^2 \tag{5.9}$$

where G_{res} = gust factor for resonant component; $= g2(\sigma_v/\overline{V})_h \sqrt{SE/\zeta}$
 M = mean base overturning moment; for a square building, it can be approxi-
        mated by $0.6\frac{1}{2}\rho\overline{V}_h^2\,bh^2$
 M_I = inertial base bending moment for unit displacement at top of building; for
        constant density and linear mode shape, $= \frac{1}{3}\rho\,bdh^2\,(2\pi n_0)^2$
 g = peak factor; for normally distributed process, $= \sqrt{2\ln Tn}$
 n_0 = first-bending-mode natural frequency; can be approximated by $46/h$,
        where h is height in meters
 $(\sigma_v/\overline{V})_h$ = longitudinal turbulence intensity at height h
 T = period under consideration, sec; usually 600 sec for acceleration criteria
 h = height of building
 b = width of building
 d = depth of building
 $\overline{V}_h$ = hourly mean wind speed at height h
 S = size factor; $= 1/[(1 + 3.5n_0h/\overline{V}_h)(1 + 4n_0b/\overline{V}_h)]$
 E = longitudinal turbulence spectrum; $= 0.47N/(2+N^2)^{5/6}$
 N = reduced frequency; $= nL_h/\overline{V}_h$
 L_h = measure of turbulence length scale; $= 1000\,(h/10)^{0.25}$
 ρ = air density
 ρ_S = building density
 ζ = critical damping ratio

Cross-Wind Response.   One of the simplest ways of evaluating the cross-wind response, involving all the important parameters in the process of resonant response to
wake excitation, is to use a mode-generalized force spectrum approach proposed by

Saunders and Melbourne (1975). The method makes use of measured cross-wind displacement spectra to give a mode-generalized force spectrum (for the first mode) of

$$S_F(n) = \frac{(2\pi n_0)^4 m^2 S_y(n)}{H^2(n)} \tag{5.10}$$

where $S_y(n)$ = spectrum of cross-wind displacement at top of building
 n_0 = first-mode frequency
 m = modal mass
 $H^2(n)$ = mechanical admittance; = $1/\{[1 - (n/n_0)^2]^2 + 4\zeta^2(n/n_0)^2\}$
 ζ = critical damping ratio

For a linear mode, and if excitation by low frequencies is small and the structural damping low so that the excitation bandwidth is large compared with the resonant bandwidth, the standard deviation of displacement at the top of the building may be approximated by

$$\sigma_y = \left| \frac{\pi n_0 S_F(n)}{(2\pi n_0)^4 m^2 4\zeta} \right|^{1/2} \tag{5.11}$$

and the standard deviation of acceleration is given by

$$\sigma_{\ddot{y}} = \sigma_y (2\pi n_0)^2 \tag{5.12}$$

The force spectrum may be expressed in coefficient form by

$$C_{FS} = \frac{n_0 S_F(n)}{(\frac{1}{2}\rho \overline{V}_h^2 bh)^2} \tag{5.13}$$

where h = building height
 b = building width normal to wind direction
 V_h = mean wind speed at top of building

Then in terms of this force spectrum coefficient the standard deviation of acceleration becomes

$$\sigma_{\ddot{y}} = \frac{\rho \overline{V}_h^2 bh}{4m} \sqrt{\frac{\pi C_{FS}}{\zeta}} \tag{5.14}$$

For an average building density ρ_S and a linear mode, the modal mass is

$$m = \frac{1}{3}\rho_S bdh \tag{5.15}$$

and the peak acceleration at the top of the building due to cross-wind response is given by

$$\hat{\ddot{y}} = \frac{3}{4} \frac{g\rho \overline{V}_h^2}{\rho_S d} \sqrt{\frac{\pi C_{FS}}{\zeta}} \tag{5.16}$$

Typical values of mode-generalized cross-wind force spectrum coefficients for a fundamental mode of vibration that has a linear mode are given in Fig. 5.4. Extension of these data to nonlinear mode shapes may be made conservatively by multiplying by a mode-shape correction factor for acceleration of $(0.76 + 0.24k)$, as discussed in

Holmes (1987), where k is the mode-shape power exponent from the representation of the fundamental mode shape by $\psi_{(\hat{z})} = (z/h)^k$.

4 Parameter Sensitivity

There are several steps to examining parameter sensitivity. First it is important to demonstrate that along-wind response is a relatively minor problem compared to cross-wind response. Second it has to be shown that mode shape is important and that it is here that the structural system can play a significant part. Third the real problem of cross-wind response has to be demonstrated along with its attendant parameter sensitivity.

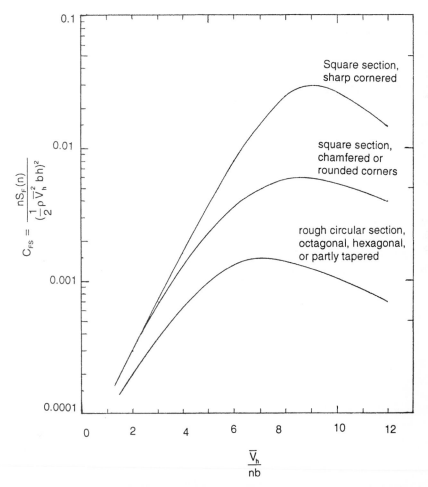

Fig. 5.4   **Typical mode-generalized cross-wind force spectrum coefficients for isolated buildings with aspect ratio $h/b = 4$ to 8 in suburban roughness for $\zeta = 0.01$.**

Worked Example.   The simplest way to introduce a study of the relative significance of the response type and the various parameters is by means of a worked example. For this purpose consider a tall building for a return period of 1 year for which

$$h = 300 \text{ m}$$

$$b = d = 50 \text{ m}$$

$$\overline{V}_h = 26 \text{ m/sec}$$

$$\rho_S = 200 \text{ kg/m}^3$$

$$(\sigma_v / \overline{V})_h = 0.12$$

$$\zeta = 0.01 \text{ (at serviceability levels)}$$

and frequency at the first mode,

$$n_0 = \frac{46}{h} = 0.15 \text{ Hz}$$

reduced velocity,

$$V_R = \frac{\overline{V}_h}{n_0 b} = 3.47$$

peak factor (10 min),

$$g = \sqrt{2 \ln 600 n_0} = 3.0$$

1. *Along-wind response.*   Gust factor for resonant component,

$$G_{res} = g \, 2 \, \frac{\sigma_v}{\overline{V}} \sqrt{\frac{SE}{\zeta}}$$

$$= 0.53$$

Mean base-overturning moment,

$$\overline{M} = 0.6 \frac{1}{2} \rho \overline{V}_h^2 b h^2$$

$$= 1.095 \times 10^9 \text{ N-m}$$

Inertial base-bending moment for a linear mode for unit displacement at the top,

$$M_I = \frac{1}{3} \rho_S b d h^2 \, (2\pi n_0)^2$$

$$= 13.32 \times 10^9 \text{ N-m}$$

Peak acceleration at the top of the building due to along-wind resonant response for a linear mode,

$$\hat{\ddot{x}} = G_{res} \frac{\overline{M}}{M_I} (2\pi n_0)^2$$

$$= 0.039 \text{ m/sec}^2$$

$$= 3.9 \text{ mg}$$

2. *Cross-wind response.*   From Fig. 5.4,

$$C_{FS} = 0.0015$$

Peak acceleration at the top of the building due to cross-wind resonant response for a cantilever mode shape, where $k = 1.5$,

$$\hat{y} = \frac{3}{4} \frac{g\rho \overline{V}_h^2}{\rho_s d} \sqrt{\frac{\pi C_{FS}}{\zeta}} (0.76 + 0.25k)$$

$$= 0.14 \text{ m/sec}^2$$

$$= 14.3 \text{ mg}$$

and for a linear mode, $k = 1.0$,

$$\hat{y} = 0.125 \text{ m/sec}$$

$$= 12.8 \text{ mg}$$

It is noted that the acceleration criterion for occupancy comfort for the 1-year return period and first-mode frequency of 0.15 Hz for 10 min is obtained from

$$\hat{y} = \sqrt{2 \ln nT} \left(0.68 + \frac{\ln R}{5} \right) \exp(-3.65 - 0.41 \ln n)$$

$$= 11.8 \text{ mg}$$

For this worked example the along-wind acceleration is well inside this criterion, but the cross-wind acceleration, even with a linear mode, is above this criterion.

Discussion of Parameter Sensitivity.

1. *Along wind versus cross wind.*   In Fig. 5.2 it was shown diagrammatically why the cross-wind accelerations dominate the problem of occupancy comfort, but the worked example shows that even for a reduced velocity of 3.47 the along-wind acceleration is about 30% of the cross-wind acceleration (3.9 versus 12.8 mg).
2. *Mode shape.*   Adjustments of the mode shape in order to get nearer a linear mode shape, by using structural systems such as k bracing at some levels to get facade columns to contribute more to resisting motion, can make a significant difference. In the worked example, going from a cantilever mode shape, $k = 1.5$, to a linear mode shape, $k = 1.0$, reduced the peak acceleration by 10% (from 14.3 to 12.8 mg). For a building on a reducing core or tube system, only with $k = 2.0$, for example, the penalty relative to a linear mode shape is around 25%.
3. *Damping.*   The cross-wind acceleration is approximately inversely dependent on the square root of the damping. It is approximate because there is a damping contribution from aerodynamic damping, which is normally positive and which then reduces the structural damping dependence. In this worked example, if the building had a reinforced concrete structural system and damping at serviceability levels of $\zeta = 0.015$, the cross-wind peak acceleration for a linear mode would become

$$\hat{y} = 12.8 \times 1.5^{-0.5}$$

$$= 10.5 \text{ mg}$$

which would bring the acceleration to within the occupancy comfort criterion of 11.8 mg.

4. *Frequency, building density, height and width, and planform shape.* The dependence of cross-wind acceleration on parameters which affect frequency and modal mass is quite complex and has been discussed and evaluated in some detail by Melbourne and Cheung (1988). The complication is mainly caused by the fact that C_{FS} is very sensitive to planform shape and reduced velocity, as shown in Fig. 5.4, and anything which impacts on frequency similarly affects reduced velocity on C_{FS}. Examples of the sensitivity of cross-wind acceleration to building height, aspect ratio, and planform shape were given in Melbourne and Cheung (1988) and are reproduced here as Fig. 5.5. From this study the overall conclusions with respect to parameter sensitivity effects on cross-wind accelerations were as follows:

    a. The acceleration is not, as one might intuitively think, dependent directly on height or aspect ratio h/d, but rather on building platform size. Indirectly height is involved because the wind speed is a function of height. Hence relatively slender buildings will have higher accelerations than squat buildings, but the important parameters here are platform size and average density—in other words, massiveness.

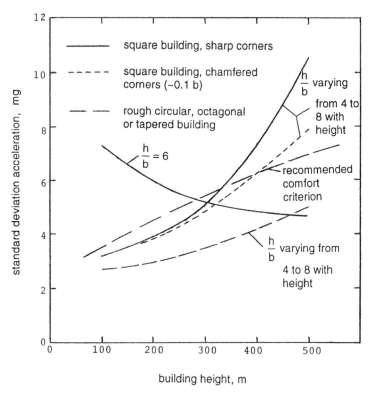

Fig. 5.5   Maximum standard deviation acceleration for 10 min in 5-year return period for various configurations; $\zeta = 0.01$; $\rho_S = 160 \ kg/m^3$; $\overline{V}_h = 22 \ (h/100)^{0.21}$; $n = 46/h$.

b. Acceleration is proportional to the square root of the force spectrum coefficient C_{FS}, and this is where parameter dependence becomes complicated. With reference to Fig. 5.4 it can be noted that C_{FS}, for a given building geometry, is expressed as a function of reduced velocity $V_R = V_h/n_0 b$ and that C_{FS} increases with V_R up to a peak; this range covers most applications. This implies an additional direct dependence on wind speed, which makes the acceleration dependent on something approaching V^3 over this region. Also the increased size described by building width b reduces V_R, and hence C_{FS}, which also works to reduce acceleration in addition to the massiveness effect. However, this size increase also moves to reduce frequency and hence increases V_R, and also C_{FS} and acceleration.

c. Modest rounding or chamfering of corners (10% of width) does not significantly reduce serviceability acceleration levels, although a significant reduction in ultimate limit-state moments can be achieved. More significant corner rounding or chamfering, so that the planform shape approaches that of a rough circle (octagon or hexagon), or some tapering with height can have a very significant effect in reducing serviceability acceleration levels. For tall buildings a 50% reduction in acceleration relative to that for a square, sharp-cornered building is reasonably achievable.

Overall the effects of frequency, building density, height and width, and planform shape are so interrelated that it is only by the type of evaluation shown in Fig. 5.5 that an appreciation of these aspects can be obtained.

5 Conclusions

The excitation mechanisms which cause the most perceptible motions in tall buildings have been described, and it has been shown that the cross-wind response is the dominant cause of motion perception problems.

Acceleration criteria for occupancy comfort have been summarized and methods of making critical estimates of acceleration levels in tall buildings have been given. Finally a parameter sensitivity discussion, with worked examples, has been presented to give a designer some indication of how to avoid high acceleration levels in tall buildings, and so avoid the need for auxiliary damping systems. In particular it was shown that very tall buildings are not necessarily the most sensitive in terms of occupancy comfort, but that square, sharp-cornered, high-aspect-ratio tall buildings are likely to have acceleration problems and that these can be avoided by using planform shapes with cut corners approaching a circular shape, tapering with height, increased mass, and structural systems which straighten up the first-mode shape.

5.2 FIRE PROTECTION OF STRUCTURAL ELEMENTS

The structural system adopted for a building, including the choice of construction materials, is often strongly influenced by the fire resistance requirements of building regulations and codes. Although building code requirements with respect to fire vary between countries, it is generally accepted that buildings should be designed for the limit state of fire to achieve the following objectives:

1. Provide an acceptable level of safety for the building occupants and firefighters.
2. The adjacent property is not damaged.

The level of safety offered to the occupants of a building in the event of a fire is a complex function of numerous factors, including:

1. The likely characteristics of the fire
2. The likely behavior of the occupants (whether they are alert or asleep, their reactions)
3. The likely performance of compartmentation with respect to restricting the movement of smoke and flames
4. The likely performance of early warning systems (if any) in notifying the occupants
5. The performance of the sprinkler system and smoke control systems (if any)
6. The response of the fire brigade

All of these factors are probabilistic by nature and functions of time. Time is of the utmost importance in designing buildings for fire safety—it being important that successful egress be achieved before conditions become untenable in the fire compartment. A systematic approach to designing buildings for fire safety needs to take into account all of these factors from a probabilistic approach and to recognize the importance of time.

In contrast to such an approach, the regulatory requirements with respect to fire safety that have evolved in many countries generally represent an ad hoc and unsystematic approach to designing buildings for fire safety. Buildings are required to be designed such that the structural members possess a certain fire resistance as determined in accordance with the standard fire test—a test that generally bears little relationship to real fires and takes no account of the time for fire development and spread. But it is a useful test in that it allows the fire resistance of elements of construction to be rated on a relative basis. Little account is taken of the types of activities taking place within the building, and generally little provision is made for the reduction of fire resistance requirements due to the presence of other components of the fire safety system such as sprinklers, smoke detectors, and more efficient egress provisions.

However, it is likely that in many situations the application of a systematic approach to assessing the fire safety of buildings will allow a substantial reduction in the level of the fire resistance required for members—without resulting in any decrease in fire safety. The purpose of this section is to consider how the structural form of buildings may be influenced by the need to design for fire safety. For a thorough consideration of fire safety in tall buildings, see *Fire Safety in Tall Buildings* (CTBUH, 1992).

At the outset it needs to be stated that concrete-framed buildings are relatively unaffected by requirements for structural members to have a level of fire resistance. This is because the fire resistance of concrete members is usually relatively easily achieved by selecting an appropriate level of cover to the reinforcement and a minimum size of member. For steel structures, on the other hand, requirements for members to have higher levels of fire resistance generally mean that members must be protected with fire-protective coverings such as sprayed insulation materials or board protection, and this can result in substantially increased costs for the steel frame. It follows therefore that it is only in the case of steel-framed buildings that there are real benefits to be gained by reducing or eliminating the need for fire protection of the structural frame. As Iyengar (1992) has stated,

It is the requirements for structural fire protection (and corrosion protection) that have inhibited the use of expressed or visible external steelwork with tall buildings. Cladding and curtain wall systems have evolved and have been used to camouflage the fire-protected steel. As the need for taller buildings has grown, it has become more important to utilize the exterior of the buildings for lateral load resistance. Unique systems such as the braced tube of the John Hancock Center, Chicago, the framed tube of the World Trade Center, New York, and the bundled tube system of the Sears Tower, Chicago, have evolved. Yet in all of these cases the external members had to be fireproofed and clad even though some structural *representation* on the facades has been achieved.

In the following, developments arising from the need to design buildings for fire safety economically and the effect of this on the choice of structural system and form of member construction are reviewed. These developments vary from innovative ways for designing steel members to achieve the specified levels of standard fire resistance as given in the building regulations, to designing members for "real" fire scenarios, to a rational engineering approach to designing for fire safety which takes into account all components of the fire safety system.

1 Design of Building Structures to Satisfy Building Code Requirements

Over the years various innovative approaches have been developed in an attempt to reduce or eliminate the need for conventional fire-protective coatings for steel members, while at the same time satisfying the (usually high) levels of standard fire resistance required by the relevant building code.

Water-Filled Members.   Around 40 buildings (IISI, 1993) have been constructed with tubular columns filled with water and with an appropriately designed circulation system to ensure that local overheating of the column does not occur and that there is a sufficient supply of water to absorb the energy associated with the required level of fire resistance. A detailed design method has been available for many years (Bond, 1975). The 64-story U.S. Steel Corporation headquarters in Pittsburgh incorporates water-filled exterior columns and is one of the tallest buildings in the world where this system has been used for providing the required fire resistance for the columns. Water cooling is most suitable for columns, although with the addition of water pumps to provide adequate circulation, the method can be used for tubular beams. For tall buildings the columns must be divided into zones to limit the buildup of pressure within the column. In general, it is true to say that the use of water filling to achieve the required standard of fire resistance for members has potential when exposed tubular steelwork is required from an architectural viewpoint.

Columns of Mixed Concrete and Steel.   The range of composite steel and concrete columns shown in Fig. 5.6 has also been used widely to provide an alternative to steel columns coated with fire-protective coatings. Both the encased I sections and the concrete-filled tubular sections offer significant advantages with respect to rapid construction. Tubular columns of large cross section have been used for tall buildings (McBean, 1990; Wyett and Bennetts, 1987; Watson and O'Brien, 1990) (see Fig. 5.7). The location of reinforcement in these members sometimes presents difficulties, and the use of unreinforced concrete is often possible, depending on the stockiness of the column, the level of load applied to the column, and the eccentricity of load. The design of mixed concrete and steel members for fire resistance is the subject of numerous publications (O'Meagher et al., 1993; British Steel, 1992; Kruppa et al., 1990; ECCS, 1988; CTBUH, 1992).

Fire-Resistant Steels.   Alternative "fire-resistant" steels have been developed (Maruoka et al., 1992; Assefpour-Dezfuly et al., 1990; CTBUH, 1992) and promoted by various steel companies, particularly from Japan. These steels give somewhat superior mechanical properties under elevated temperature conditions compared with conventional steels, although use of these steels will not remove the necessity for a fire-protective coating—a lesser thickness of fire protection will need to be applied and the steels are generally more expensive than conventional steels.

2 Design of Building Structures for "Real" Fire Scenarios

"Real" Fires versus Standard Fires.   The previous section has dealt with the design of buildings where the members are required to have levels of fire resistance as determined in accordance with the standard fire test (ISO, 1985). The time-temperature curve associated with the standard fire test varies markedly compared with those associated with real fires (Fig. 5.8). This matter will not be considered in detail here except to note that this has been demonstrated by fire tests that have been conducted in various-size compartments having different surface linings, various quantities of fuel (normally represented by timber and plastic cribs), and varying degrees of ventilation (Pettersson et al., 1976). Other fire tests have been conducted with real furniture in small and, more recently, in large fire compartments (Thomas et al., 1992a). Based on compartment tests conducted in room-size enclosures with the fire load represented by cribs, various engineering models have been developed to predict the temperature (and time-temperature)

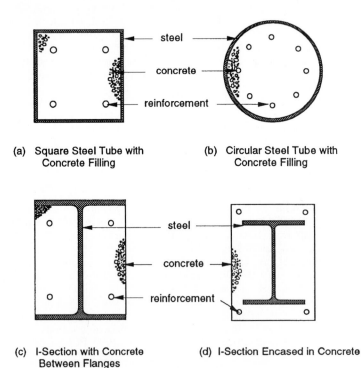

(a) Square Steel Tube with   (b) Circular Steel Tube with
     Concrete Filling   Concrete Filling

(c) I-Section with Concrete   (d) I-Section Encased in Concrete
     Between Flanges
     (Arbed Column)

Fig. 5.6   Types of composite column sections.

Fig. 5.7   Forrest Centre, Perth, Australia.

conditions given a certain level of ventilation and fire load (Pettersson et al., 1976). Through such testing it has been recognized that under certain conditions, it is possible to reduce (or even eliminate) the level of fire protection required for structural members.

Buildings with External Steelwork.   It has been shown (Law and O'Brien, 1981; Kruppa, 1981) that the location of steelwork beyond or at the facade of the building, or such that it is partly shielded from flames which may come from the windows in the event of a fire, will under certain ventilation conditions result in temperatures that are not sufficiently high to require fire protection of the steelwork. Temperatures experienced at (or beyond) the facade are generally considerably lower than those within the fire compartment. This fact has been demonstrated by means of fire tests in compartments where the fire load has been generally represented by wood cribs and the fire compartments have various degrees of ventilation.

This approach has resulted in the use of unprotected external steelwork in numerous buildings such as Bush Lane House, London (Fig. 5.9) (Brozzetti et al., 1983), where the steelwork forming the external lattice is of relatively small cross section and cooled by water.

The Hotel de las Artes tower in Barcelona, Spain (Fig. 5.10), is a very recent example of the use of unprotected external steelwork (Iyengar, 1992). In this case the outer columns and the lateral bracing system are located outside the building facade. Calculations were performed using the method given by Law and O'Brien (1981), assuming a given fire load in a hotel compartment and a representative level of ventilation. The calculated temperatures for the external steelwork were confirmed by means of a fire

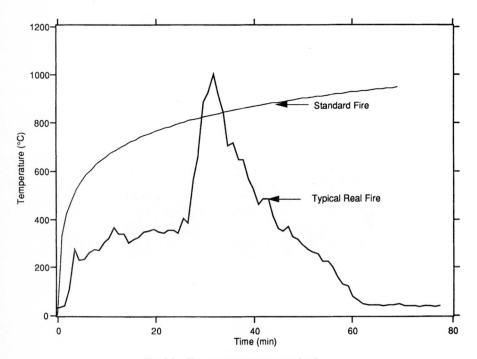

Fig. 5.8   Time-temperature curves for fires.

Fig. 5.9   Example of use of water-filled tubes; Bush Lane House, London, U.K.

Fig. 5.10   Hotel de las Artes, Barcelona, Spain.

test. It is clear that in this case the regulatory authorities were prepared to accept this approach in lieu of all members having to achieve the higher level of fire resistance required by the regulations.

Similar calculations have been used in Japan (Sakumoto et al., 1992) for high-rise buildings to permit the use of unprotected "fire-resistant" steel at the facade.

Parking Garages.   The fire load and ventilation conditions associated with parking garages are well known. Open-deck parking garages are generally defined as buildings that have at least two opposite sides open to at least 50%. Fire tests involving cars in open-deck situations have been conducted in the United Kingdom, Japan, the United States, and Australia (IISI, 1993; Bennetts et al., 1985). These tests have shown that, provided the structural members are at least of a certain size—and this size is met with practical sections used in parking garages—the temperatures achieved will not lead to off-loading of the structural members. Thus multistory steel parking garages without fire protection cladding are permitted in many countries of the world.

Tests have been conducted on closed parking garages and those which are partially open but do not comply with the preceding definition of open deck (Bennetts et al., 1989). The tests showed that the fire temperatures in partially open parking garages can be equivalent to those that would be experienced in a closed garage, which in turn are higher than those that will be achieved in open-deck garages. In Australia, for garages containing more than 40 cars, those that are not open-deck are required to be sprinklered due to the potential difficulties in extinguishing fires in these garages due to the dense smoke. The building regulations in Australia now allow the construction of closed parking garages with unprotected structural steel.

Mixed-Occupancy Buildings.   The matter of mixed-occupancy buildings is now considered. Multistory buildings often incorporate stories which under the building regulations are required to have a more fire-resistant type of construction or a higher level of fire resistance than the stories below. Many building regulations then require that the lower stories have the same type of construction and level of fire resistance as the higher stories. For example, in many countries where isolated open-deck parking garages are permitted to be constructed in unprotected steel, this would not be permitted if the open-deck garage formed the lower levels of an office building.

Arguments have been advanced and accepted in Australia (Thomas et al., 1989) where numerous buildings have now been permitted to be constructed with unprotected steel parking levels below stories of offices and shops. Figures 5.11 and 5.12 show one such example, where four levels of open-deck parking garage constructed from non-fireproofed steelwork are located below 12 stories of office accommodation.

Essential and Nonessential Members.   Building codes usually require all members within a fire compartment to have the same level of fire resistance, irrespective of whether under fire conditions, the presence of the member is necessary to preserve compartmentation or structural adequacy of the building. This should not be the case, as has been successfully argued in a number of situations. For example, the building shown in Fig. 5.13 is a high-rise building incorporating large-diameter concrete-filled tubes, composite concrete floors, and a reinforced concrete service shaft. External trusses spanning between the columns were provided to ensure adequate lateral load resistance under design ultimate wind forces and adequate lateral stiffness under service wind loads. The architect required the external bracing to be of exposed steelwork, yet under the relevant building regulations these members were required to achieve a fire resistance of 120 min. Nevertheless the case for unprotected steel trusses was successfully argued on the basis that in the extreme event of fire, the presence of the bracing was not necessary to provide lateral support to the columns or resist any wind loading likely to be applied to the building during the fire. These functions were fulfilled by the presence of the braced service core. The wind load considered for the fire situation should be very much lower than that adopted for the purpose of normal temperature design. This is because the likelihood of getting a combination of extreme values of load and fire (itself an extreme event) is very small.

This approach can be extended to apply to a number of situations. It is clearly necessary to identify which members are necessary to provide overall building stability and to maintain compartmentation. These members should be designed to have the required level of fire resistance; the remaining members do not require fire protection.

3 Systematic Design of Buildings for Fire Safety

As discussed, the primary purpose of the requirements for the fire protection of structural members, and other elements within buildings, is to achieve a satisfactory level of fire safety. Nevertheless, the level of fire safety offered by a building is often, in reality, more influenced by the other components of the fire safety system than by the level of fire resistance required for the structure. The conventional regulatory approach does not generally consider the possibility of reducing the requirements of one or more components of the fire safety system (such as the level of fire resistance required) given the presence of other components, such as the incorporation of a reliable sprinkler system.

The development of a rational approach which recognizes the influence of all components of the fire safety system may have an important effect on the choice of structural system. It has been generally accepted that this may be achieved by a method which rationally determines the risk to the occupants and the effectiveness of the components of a fire safety system in a building.

Fig. 5.11   Open-deck garage; 624 Bourke Street, Melbourne, Australia.

Over the last few years considerable progress has been made in developing, documenting, and establishing the necessary acceptance for a method of risk assessment which is capable of meeting these needs and of forming the basis for a systematic procedure for the analysis and design of fire safety in buildings (CTBUH, 1980; Beck, 1991; Thomas et al., 1992b). An outcome of this work has been the realization that existing requirements for levels of fire resistance in the building regulations are often excessive, and that there are more economical and appropriate means of achieving the required levels of fire safety in buildings.

Fig. 5.12   624 Bourke Street.

Fig. 5.13   Proposed office building, Sydney, Australia.

An illustration of the benefits that may be achieved by the approach described is illustrated by a research program undertaken to investigate options associated with the refurbishment of a 41-story building. The building, shown in Fig. 5.14, incorporates a braced steel core and closely spaced exterior steel columns which combine with steel spandrel beams to form an exterior tube structure. The K-braced core is connected to the exterior tube by means of transfer trusses at the top and midheight of the building. Belt trusses extending around the perimeter of the building are located at the top, midheight, and bottom of the building. All of these steel members are fire protected by means of concrete encasement, which in the case of the exterior columns, is further encapsulated

Fig. 5.14   140 William Street, Melbourne, Australia.

by steel plate. In addition, the core is separated from the rest of the area of each story by masonry walls. The floor beams and composite floor slabs are protected with asbestos-based fire protection material as is the inside surface of the facade beams.

The sprinkler system in the building does not comply with current code requirements for sprinkler head spacing or water delivery rates. Moreover, no sprinklers are located in the ceiling space, as is required for current construction.

The proposed refurbishment of the building required the removal of asbestos-based fire protection material from the beams which support the floor slabs and from the soffit of the composite floor slabs. For the refurbished building to meet the deemed-to-comply requirements of the regulations, it would require respraying of the beams and floor slab soffit, alteration of the sprinkler system to change it from an extralight hazard system to an ordinary hazard system, and the fitting of sprinklers in the ceiling spaces. In contrast, the building owner proposed that the refurbished building retain the existing sprinkler heads and that the slabs and floor beams remain unprotected. The reliability of the sprinkler system was further improved by the inclusion of additional monitored valves and a system to enable weekly checking of the presence of water in the sprinkler pipes at every floor.

At the request of the building owner, a series of fire tests and a risk assessment were undertaken. The risk assessment was conducted by systematically modeling the events that might follow the occurrence of a fire in the building, and by using a Monte Carlo simulation to evaluate the probability of outcomes which would lead to deaths among the occupants of the building. The risk assessment was carried out for two (conceptual) situations—the building designed to satisfy all of the minimum requirements of the current building regulations; and the proposed refurbished building as described. Each of the models of the building accurately accounted for the layout of the building and the subsystems and components of the fire safety system. Many of the data on fire growth and development, smoke movement, and alarm cues required for the risk assessment came from an extensive test program (Thomas et al., 1992a) in which four fire tests were conducted in a test building specially constructed to simulate part of the prototype building.

The results of the risk assessment showed that the risk to life safety in both buildings is low, but that the refurbished building is substantially safer than that satisfying the minimum requirements of the regulations. On the basis of these findings the building has been refurbished such that the existing sprinklers remain and no fire protection is applied to the steel beams or floor slabs.

Further testing and research is being undertaken to provide the basis for a more generalized approach to determining the level of fire safety offered by a building based on a rational consideration of the factors described earlier. Clearly such an approach has the potential to offer substantial flexibility with respect to structural form, as the influence of all components of the fire safety system can be taken into account.

5.3 CONDENSED REFERENCES/BIBLIOGRAPHY

Assefpour-Dezfuly 1990, *Fire Resistant High Strength Low Alloy Steels*

Beck 1991, *Fire Safety Systems Design Using Risk Assessment Models—Developments in Australia*

Bennetts 1985, *Open-Deck Carpark Fire Tests*

Bennetts 1989, *Fire in Carparks*

Bond 1975, *Fire and Steel Construction: Water Cooled Hollow Columns*

British Steel 1992, *Design Manual for Concrete Filled Columns*

Brozetti 1983, *Fire Protection of Steel Structures—Examples of Applications*

Chen 1973, *Human Perception Thresholds to Horizontal Motion*

CTBUH Group CL 1980, *Tall Building Criteria and Loading*

CTBUH Committee 8A 1992, *Fire Safety in Tall Buildings*

Davenport 1967, *Gust Loading Factors*

ECCS 1988, *Calculation of the Fire Resistance of Centrally Loaded Composite Steel-Concrete Columns Exposed to the Standard Fire*

Holmes 1987, *Mode Shape Corrections for Dynamic Response to Wind*

IISI 1993, *Fire Engineering Design for Steel Structures: State of the Art*

Irwin 1986, *Motion in Tall Buildings*

ISO 1985, *Fire-Resistance Tests—Elements of Building Construction*

Iyengar 1992, *Hotel de las Artes Tower, Barcelona, Spain*

Kruppa 1981, *Fire-Resistance of External Steel Columns*

Kruppa 1990, *Structural Fire Design*

Law 1981, *Fire Safety of External Steelwork*

Maruoka 1992, *Development and Test Results of SM520B-NFR Fire Resistant Steel for Procter & Gamble Far East, Inc. Japan Headquarters Building*

McBean 1990, *The MYER Centre, Adelaide—A Case Study*

Melbourne 1977, *Probability Distributions Associated with the Wind Loading of Structures*

Melbourne 1980, *Notes and Recommendations on Acceleration Criteria for Occupancy Comfort in Tall Structures*

Melbourne 1988, *Designing for Serviceable Accelerations in Tall Buildings*

Melbourne 1992, *Accelerations and Comfort Criteria for Buildings*

O'Meagher 1993, *Behaviour of Composite Columns in Fire*

Pettersson 1976, *Fire Engineering Design of Steel Structures*

Reed 1971, *Wind Induced Motion and Human Comfort*

Sakumoto 1992, *Application of Fire-Resistant Steel to a High-Rise Building*

Saunders 1975, *Tall Rectangular Building Response to Cross-Wind Excitation*

Thomas 1989, *Fire in Mixed Occupancy Buildings*

Thomas 1992a, *Fire Tests of the 140 William Street Office Building*

Thomas 1992b, *The Effect of Fire on 140 William Street—A Risk Assessment*

Vickery 1966, *On the Assessment of Wind Effects on Elastic Structures*

Vickery 1969, *On the Reliability of Gust Loading Factors*

Watson 1990, *Tubular Composite Columns and Their Development in Australia*

Wyett 1987, *Structural Fire Engineering in Building Design—A Case Study*

6

Systems for the Future

A look at the future not only concludes this study of state-of-the-art structures but also opens the door for another monograph in this series. The subject of unbuilt projects and future systems, a rich mix of visionary projects from around the world, has fascinating potential for further exploration and presentation in a separate volume.

This final chapter will serve as a brief summary of where tall building systems seem to be headed in the near, rather than distant, future. Several project descriptions are appended to this section, which illustrate some of these principal tendencies. The projects demonstrate the rich diversity of systems now available to designers. They are all recently designed unbuilt projects, utilizing systems discussed in earlier chapters.

- *Core and outrigger systems:*   Miglin-Beitler Tower, Chicago and Dearborn Center, Chicago
- *Trussed tube systems:*   Shimizu Super High Rise, Tokyo
- *Hybrid systems:*   Bank of the Southwest Tower, Houston

The reasons that these buildings remain unbuilt range from changing economic conditions, as in the case of the Bank of the Southwest, to projects that await financing in a slow market, such as the Shimizu Super High Rise. In addition to their unbuilt status, they also share some features that illustrate tendencies in tall building design. These include architectural, structural, as well as other tendencies that point to the future.

Before discussing the features of these towers, it is worth mentioning one visionary project, of the type that might appear in a future monograph, as suggested. It has some features common to the other schemes presented in this chapter, extrapolated to a height significantly taller. William LeMessurier has proposed a half-mile-high tower [850 m (2789 ft)], the Erewhon Center (Fig. 6.1) (Architectural Record, 1985). With a floor plan approximately the size of the Sears Tower or the World Trade Center it has usable floor areas proven in existing tall buildings. The structural systems for this tall building have more in common with the unbuilt projects of this chapter than the current record holders. The use of massive high-strength concrete columns on the exterior, cast composite with the structural steel frame, utilize the cost-effective strength and stiffness of concrete in compression. Bracing is employed both as a lateral resistance system and as a gravity load transfer system to allow all load-bearing columns to participate in the lateral resistance for optimum efficiency. The result is a very rigid tower with a 10-sec period of vibration, utilizing conventional construction techniques.

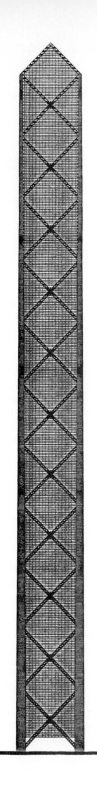

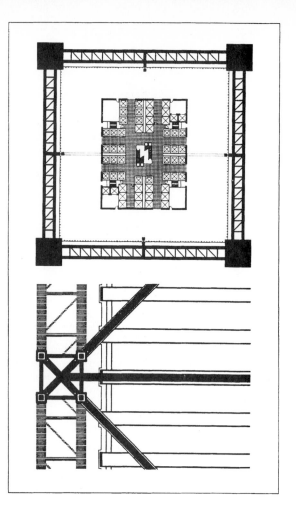

Fig. 6.1　Erewhon Center.

6.1   *ARCHITECTURAL TENDENCIES*

And so what are some of the current tendencies in tall building design that can be expected to continue in the late twentieth century and into the next? There is no single architectural trend, as in the 1960s and 1970s, that dominates the design of tall buildings. There are, of course, buildings that utilize structure as part of the architectural expression, in the tradition of projects such as Chicago's John Hancock Tower, whereas other building structures, primarily from the 1980s, defer to the architectural massing chosen more in consideration of urban design issues. Greater use of mixed systems, along with architectural trends toward utilizing the structural systems as a form generator (along with urban design), are blurring these earlier distinctions and creating many options for the 1990s and beyond.

The Bank of the Southwest Tower exhibits the potential for the massing and architectural expression to accent and reveal the structural system. The massive composite columns reduce in size with height, and the architect takes advantage of the column locations and dimensions to shape the tower in a more dynamic and soaring expression than a simple prismatic form. The Miglin-Beitler Tower and the Dearborn Center, utilizing a core and outrigger system, achieve similar forms, but with different slenderness proportions and tops.

The Shimizu Super High Rise is a trussed tube with some other similarities with Chicago's Hancock. They are both mixed-use buildings, with offices below and residential floors above. This requires smaller floor plates in upper floors. The Hancock achieves this with a constantly sloping exterior truss tube, whereas Shimizu rotates the tube, resulting in smaller floor plates with each rotation. These two projects illustrate the opportunity presented by an exterior truss geometry as the primary source of architectural expression, while at the same time adhering to an efficient and rigid structural system.

6.2   *STRUCTURAL TENDENCIES*

The systems for the tall buildings presented here all take full advantage of the mass, width, and potential efficiencies of the towers. The similarities among these building systems illustrate, by example, design ideas that work in many different conditions. The following list summarizes these common features:

- Composite elements
- Use of high-strength concrete for supercolumns
- Bracing or core walls for lateral stiffness
- Use of active and passive damping systems
- Use of better analytical tools and testing facilities

There is a greater tendency to mix systems and materials today, particularly concrete and steel. Composite steel and concrete floor systems are utilized in all of the projects that follow. In addition to efficient use of materials, the self-shoring nature of the system lends itself to the requirements for fast construction. The improvement of high-strength concrete has made possible the massive columns utilized in all of the schemes, which proved to be an economical way to achieve axial strength and stiffness and, by extension, bending stiffness.

The use of braced frames or shear walls, in lieu of moment resistant frames, is also evident. These systems are inherently more stiff, and therefore more economical in achieving drift and acceleration limits. Bracing and walls are locally more limiting than framed tubes, particularly when bracing penetrates the interior volume. But bracing and core walls also open up other opportunities for flexibility. An example of this is the exterior wall of braced towers, where columns may be much smaller than the massive sections required for framed tubes.

Another advancement in the performance of tall buildings is the use of damping systems. Active damping systems were first used in the retrofit of Boston's Hancock building, as well as an original design feature in New York's Citicorp Center. The World Trade Center was one of the first to use passive damping systems. The use of these systems is becoming more common now, and indeed the Shimizu tower proposes an active damping system (HMD). The improvement in analytical tools, namely, more powerful computers at affordable prices, has made some of these advancements possible. And improvements in testing facilities, both shaking tables and wind tunnels, have also aided the understanding and usefulness of these systems. Base isolation systems for earthquake motion, as well as tuned mass dampers for the control of wind movements, are now common design considerations. Other systems, such as active control of building structures with advanced microprocessors, are also being tested, and increasing use could be anticipated in the future.

6.3   OTHER TENDENCIES

Finally there is movement toward greater integration in the design and construction process through information systems. Consideration of construction methods and systems, including prefabrication, modular construction, and robotics, is changing the traditional project delivery systems. Information systems for monitoring quality assurance during construction as well as monitoring the long-term performance of buildings are also on the horizon, with the integration of mechanical, vertical transportation and maintenance systems.

PROJECT DESCRIPTIONS

Miglin-Beitler Tower
Chicago, Illinois, USA

Architect	Cesar Pelli Associates Inc. with HKS Inc.
Structural engineer	Thornton-Tomasetti Engineers
Year of completion	Future
Height from street to roof	610 m (2000 ft)
Number of stories	141
Number of levels below ground	1
Building use	Office
Frame material	Concrete core, major columns, outrigger walls, steel floor beams, Vierendeel trusses
Typical floor live load	2.5 kPa (50 psf)
Basic wind velocity	33 m/sec (73 mph) at 10 m (33 ft); 46 m/sec (103 mph) at 610 m (2000 ft), 50-yr return
Maximum lateral deflection	711 mm (28 in.) at 110th floor, 50-yr return
Design fundamental period	9 sec
Design acceleration	23 mg peak, 10-yr return
Design damping	1.5 to 2% serviceability
Earthquake loading	$ZC = 0.0042$; horizontal force factor 1.33
Type of structure	Concrete core linked by concrete beams to eight major perimeter concrete columns
Foundation conditions	30-m (100-ft) silty and sand clay over dolomitic limestone bedrock
Footing type	27.4-m (90-ft) deep, 2.4- to 3-m (8- to 10-ft)-diameter caissons socketed into rock
Typical floor	
Story height	3.96 m (13 ft)
Beam span	10.67 m (35 ft)
Beam depth	460 mm (18 in.)
Beam spacing	3.05 m (10 ft)
Material	Steel
Slab	89-mm (3.5-in.) normal-weight concrete on 76-mm (3-in.) metal deck
Columns	
Size at ground floor	11 by 2 m (36 by 6.5 ft)
Spacing	18.6 m (61 ft)
Material	100-MPa (14,000-psi) concrete

Core   Concrete, 100 to 70 MPa (14,000 to
                                              10,000 psi)
   Thickness at ground floor   914, 460 mm (36, 18 in.)

The structural system for the proposed 141-story 610-m (2000-ft)-high Miglin-Beitler office building has been designed by the structural engineering firm Thornton-Tomasetti Engineers of New York City (Fig. 6.2). A simple and elegant integration of building form and function has emerged from close cooperation of architectural, structural, and development team members. The resulting cruciform tube scheme offers structural efficiency, superior dynamic behavior, ease of construction, and minimal intrusion at leased office floors (Fig. 6.3).

Major objectives of the structural design were to achieve speed and economy of construction and avoid interior columns in order to maximize net rentable area. This was achieved through a structural concept based on a cruciform (crosslike) tube which, in plan, is similar in appearance to a tic-tac-toe board. The simplicity of this structural grid allows structural elements for the slender tower to continue uninterrupted from the bottom of the building to the top.

The cruciform tube structural system consists of the following six major components:

1. A 19- by 19-m (62.5- by 62.5-ft) concrete core with walls of varying thickness. The interior cross walls of the core are generally not penetrated with openings. This contributes significantly to the lateral stiffness.

2. Eight cast-in-place concrete fin columns located on the faces of the building, which extend up to 6 m (20 ft) beyond the 42.6- by 42.6-m (140- by 140-ft) tower footprint.

3. Eight link beams connecting the four corners of the core to the eight fin columns at every floor. These reinforced concrete beams are haunched at both ends for increased stiffness and reduced in depth at midspan to allow for passage of mechanical ducts. Linking the fin columns and core enables the full width of the building to act in resisting lateral forces. In addition to link beams at each floor, sets of two-story-deep outrigger walls are located at levels 16, 56, and 91. These outrigger walls enhance the interaction between exterior fin columns and the core.

4. A conventional structural steel composite floor system with 460-mm (18-in.)-deep rolled steel beams spaced at approximately 3 m (10 ft) on center. A slab of 76-mm (3-in.)-deep 1-mm (20-gauge) corrugated metal deck and 89 mm (3.5 in.) of stone concrete topping spans between the beams. The steel floor system is supported by the cast-in-place concrete elements.

5. Exterior steel Vierendeel trusses consisting of the horizontal spandrels and two vertical columns at each of the 18.6-m (61-ft)-wide faces on the four sides of the building between the fin columns. To eliminate stresses produced by creep and shrinkage strains in the concrete fin columns, the verticals in the Vierendeel are provided with vertical slip connections. This has the added benefit of channeling all of the gravity loads on each of the building faces out to the fin columns to help eliminate uplift forces on the foundations.

Exterior steel Vierendeel trusses are used to pick up each of the four cantilevered corners of the building. Corner columns are eliminated, providing for corner offices with undisturbed views. Connections between the steel Vierendeel trusses and the concrete fin columns are typically simple shear connections which minimize costs and expedite erection.

6. A 183-m (600-ft)-tall steel-framed tower at the top of the building. This braced frame is to house observation levels, window washing, mechanical equipment rooms, and an assortment of broadcasting equipment.

Fig. 6.2 Miglin-Beitler Tower, Chicago, Illinois.

A cruciform tube structure provides a safe, elegant, efficient, and constructible solution to the challenge of designing the world's tallest building, the Miglin-Beitler Tower. The proposed structural solution combines the erection speed of concrete construction, the flexibility for future change and the efficiency for horizontal spans of a steel floor system, and the superior dynamic acceleration response of a composite lateral load resisting structural system.

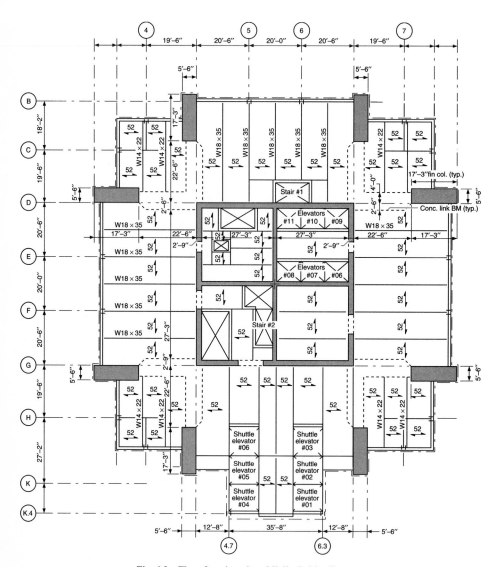

Fig. 6.3   Floor framing plan; Miglin-Beitler Tower.

Dearborn Center
Chicago, Illinois, USA

Architect	Skidmore Owings and Merrill
Structural engineer	Skidmore Owings and Merrill
Year of completion	Proposal only
Height from street to roof	346 m (1135 ft)
Number of stories	85
Number of levels below ground	3
Building use	Office
Frame material	Concrete core, steel perimeter frame, steel outrigger trusses
Typical floor live load	2.5 kPa (50 psf)
Basic wind velocity	34 m/sec (75 mph)
Maximum lateral deflection	$H/500$, 100-yr return period
Design fundamental period	7.9 sec
Design acceleration	22 mg, 10-yr return period
Design damping	1.75% serviceability
Earthquake loading	Not applicable
Type of structure	Concrete core, steel perimeter frames, steel outrigger and belt trusses
Foundation conditions	24.4 m (80 ft) of clay over bedrock
Footing type	Concrete caissons with steel liner
Typical floor	
Story height	3.96 m (13 ft)
Beam span	13.7 m (45 ft)
Beam depth	762 mm (30 in.)
Beam spacing	3.05 m (10 ft)
Material	Steel, grade 250 MPa (36 ksi)
Slab	63-mm (2.5-in.) lightweight concrete on 76-mm (3-in.) metal deck
Columns	
Size at ground floor	914 by 610 mm (36 by 24 in.)
Spacing	9.14 m (30 ft)
Material	Steel, grade 350 MPa (50 ksi)
Core	Concrete shear walls, 760 mm (30 in.) thick at ground floor; strength 49 MPa (7000 psi)

The project will consist of an equivalent 85-story office tower with a total overall gross enclosed area of approximately 246,000 m^2 (2.6 million ft^2) of which approximately 227,000 m^2 (2.4 million ft^2) is above grade (Fig. 6.4).

The first five floors will cover an area approximately equivalent to the site and will contain approximately 9270 m^2 (98,000 ft^2) of retail space on the ground floor, con-

Fig. 6.4   **Dearborn Center, Chicago, Illinois.** (*Photo by Hedrich-Blessing.*)

course level, and second floor (Fig. 6.5). The office tower will be located at the west end of the site. Figure 6.6 shows the outrigger truss system used.

There will be three below-grade levels. The concourse level contains retail rental space plus mechanical, electrical, and building services areas. The second and third lower levels will be devoted primarily to parking for 237 cars, but will also contain the main incoming electric and telephone services, employee facilities, and tenant areas.

A multilevel retail galleria will extend from the concourse level up through the second floor and will interconnect with the Dearborn Street and State Street subway stations at the concourse level. The retail levels will be linked by escalators within a skylighted, stepped atrium space. Two additional pairs of escalators will connect the first, second, and fourth floors at the elevator core. Offices spaces on the third, fourth, and fifth floors will open into the atrium.

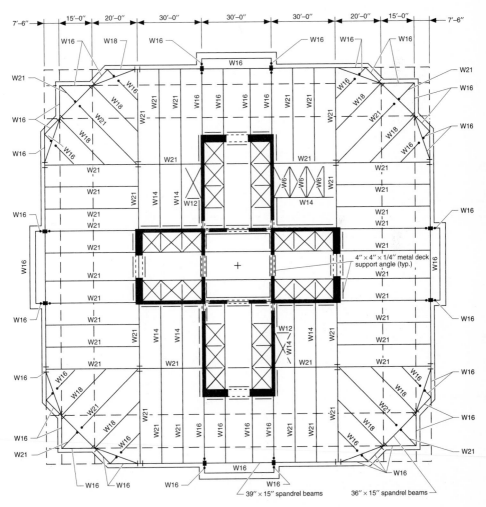

Fig. 6.5   Low-rise framing plan; Dearborn Center.

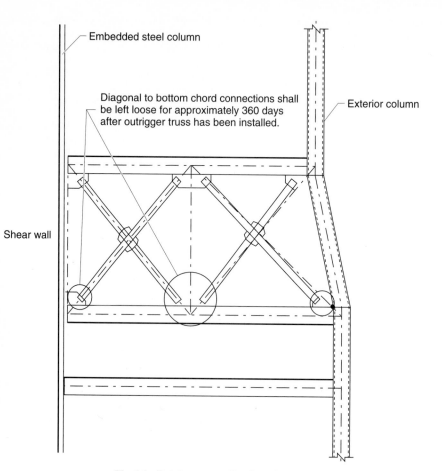

Embedded steel column

Diagonal to bottom chord connections shall be left loose for approximately 360 days after outrigger truss has been installed.

Exterior column

Shear wall

Fig. 6.6   Outrigger truss; Dearborn Center.

Bank of the Southwest Tower
Houston, Texas, USA

Architect	Murphy/Jahn with Lloyd Jones Brewer Associates
Structural engineer	LeMessurier Consultants with Walter P. Moore and Associates
Year of completion	Never built
Height from street to roof	372 m (1222 ft)
Number of stories	82
Number of levels below ground	4
Building use	Office and retail
Frame material	Steel with concrete supercolumns
Typical floor live load	2.5 kPa (50 psf)
Basic wind velocity	47 m/sec (105 mph), 100-yr return
Maximum lateral deflection	1167 mm (3.83 ft), 100-yr return
Design fundamental period	7, 6.75 sec horizontal; 7 sec torsion
Design acceleration	22 mg peak with TMD; 40 mg without
Design damping	1 to 1.2% serviceability; 3.5% with TMD; 1.5% ultimate
Earthquake loading	Not applicable
Type of structure	9-story-high A-frame trusses spanning building between concrete supercolumns
Foundation conditions	At least 76 m (250 ft) of very stiff clay
Footing type	75-m (245-ft)-wide octagonal mat, 4 to 1.8 m (13 to 6 ft) thick, 17 m (56 ft) below grade
Typical floor	
Story height	3.96 m (13 ft)
Beam span	14.2, 13.4, 11.6 m (46.75, 43.92, 37.92 ft)
Beam depth	530, 460, 410 mm (21, 18, 16 in.)
Beam spacing	3.05 m (10 ft)
Material	Steel, grade 350 MPa (50 ksi)
Slab	63-mm (2.5-in.) lightweight concrete on 50-mm (2-in.) metal deck
Columns	8 columns, 2.9 by 6 m (9.5 by 19.7 ft), tapered to 1.37 by 1.6 m (4.5 by 5.27 ft) at roof; 70-MPa (10,000-psi) concrete at base
Core	Steel, grade 250 and 350 MPa (36 and 50 ksi) supported on A-frame trusses

The tapered form of this mixed-construction 372-m (1222-ft)-high tower, its peaked sculptured crown, and the slender spire to top it off recall the dramatic upward-reaching skyscrapers of the 1930s (Fig. 6.7). The architects were chosen as a result of a design competition held by the developer. Unfortunately the Texas oil-based recession made it necessary to cancel the project after completion of the design development. The tower contained an area of over 204,400 m^2 (2.2 million ft^2). At ground level and below there were retail space and parking in addition to a grand lobby space. The tower was set diagonally on its downtown Houston site.

The tower was square with shaped corners to provide more offices with corner windows, and tapered from 55 to 46 m (180 to 150 ft) square at the eightieth floor. It rested on only eight large concrete columns, which diminished in cross section from the top of the foundation mat to floor 80 (Fig. 6.8).

The overall structural slenderness ratio of the tower was 8.0, based on 390 m/48.7 m (1279 ft/160 ft), the ratio of the tower height above the top of the mat to the horizontal dimension center to center of the columns at that level.

The severe Houston wind climate, the high slenderness ratio of the tower, and its narrow width led to wind having a controlled influence on the structural design. The structural framing system was selected to maximize structural efficiency for lateral strength and stiffness, with the least cost premium over that required for gravity loads and minimal interference with architectural layout. The main structural frames were four steel supertrusses, two in each direction, which carry the entire building load out to the concrete columns. The supertrusses had diagonals in a chevron pattern at nine-story intervals, with horizontal ties at the fourth and ninth story of each. The diagonals only appeared outside the central service core for four stories out of each nine-story module (Fig. 6.9).

The entire floor space outside the core was otherwise column-free, with conventional composite steel beams spanning from the core to a perimeter steel girder. The 24.4-m (80-ft)-wide core was bridged by a pair of Vierendeel trusses. The eight high-strength concrete columns contained embedded steel erection columns as well as reinforcing bars.

Extensive wind tunnel testing at the University of Western Ontario indicated that because of vortex shedding, the tower would have excessive wind forces and lateral accelerations unless its vibration period was limited to above 7 sec, a low value for so tall a structure. Even at that period, the tower occupants would experience too frequent discomfort from wind-induced motion. A special study was made to assess the amount of additional damping that the foundation-mat–soil interaction would provide (approximately 0.3%). In order to reduce accelerations to acceptable levels, a tuned mass damper system, of a type similar to that installed in New York's Citicorp Center, was to be located in the crown of the tower at 352.6 m (1157 ft) above ground. The mass block was to have a weight of about 386 tonnes (425 tons) and was designed to increase the tower effective damping to at least 3.5%.

Fig. 6.7   Bank of the Southwest Tower, Houston, Texas.

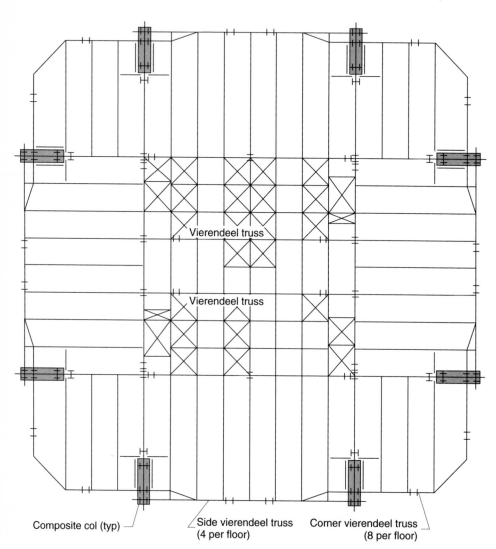

Fig. 6.8   Floor plan; Bank of the Southwest Tower.

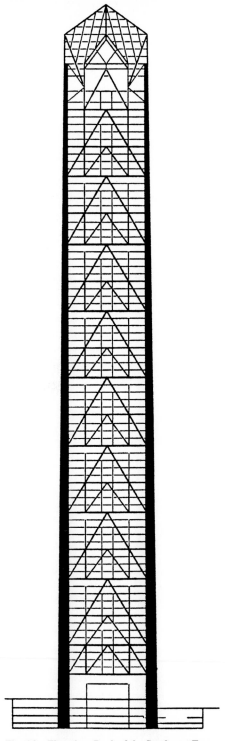

Fig. 6.9   Elevation; Bank of the Southwest Tower.

Shimizu Super High Rise (SSH)
Tokyo, Japan

Architect	Shimizu Corporation
Structural engineer	Shimizu Corporation
Year of completion	Proposal
Height from street to roof	550 m (1804 ft)
Number of stories	121
Number of levels below ground	6
Building use	Hotel, offices, retail shops, halls, parking
Frame material	Steel reinforced concrete
Typical floor live load	1.8, 3 kPa (36, 60 psf)
Basic wind velocity	45.5 m/sec (102 mph)
Maximum lateral deflection	$H/300$ (level 1 loading); $H/200$ (level 2 loading)
Design fundamental period	6.0 sec
Design acceleration	5 mg peak, 1-yr return
Design damping	0.6% serviceability; 2% ultimate
Earthquake loading	Seismic response factor 0.05
Type of structure	Trussed tube megastructure
Foundation conditions	160 m (525 ft) of sand
Footing type	Combination of continuous walls and pre-tensioned high-strength concrete (PHC) piles
Typical floor	
Story height	3.25 m (10 ft 8 in.) hotel; 4.3 m (14 ft 1 in.) office
Beam span	22.4, 15.8 m (73 ft 6 in., 51 ft 10 in.)
Beam depth	1.2, 0.9 m (47, 35 in.)
Beam spacing	12.8, 20.0 m (42 ft, 65 ft 7 in.)
Material	Steel
Slab	U-type steel deck + lightweight concrete, 155 mm (6 in.) thick
Columns	
Size at ground floor	4.0 by 2.4 m (13 by 8 ft)
Spacing	26.0, 12.8 m (85, 42 ft)
Material	Steel and concrete; HT60, $F_c = 60$ MPa (8500 psi)
Core	Braced frame
Material	Steel, HT60
Thickness	1.2 m (47 in.)

The SSH building is 550 m (1804 ft) tall with 121 stories above ground and six stories underground (Fig. 6.10). This design project was intended to confirm the feasibility of constructing such a tall building in the earthquake- and typhoon-prone country of Japan by the end of this century based on the technologies available today at Shimizu.

The SSH building was designed as a complex consisting of hotels, offices, and shops. The building area is 44,000 m^2 (474,000 ft^2) for a plot area of 90,000 m^2 (969,000 ft^2). The total space of the SSH building is 754,000 m^2 (8,116,800 ft^2) and is divided into three zones along the height. A zone was designed to be squeezed through the top and

Fig. 6.10   Shimizu Super High Rise (SSH), Tokyo, Japan.

rotated by 45° against the lower zone. The bottom zone, zone 1, consists of 43 stories with an average floor space of about 6200 m² (66,700 ft²). The middle zone, zone 2, consists of 37 stories with an average floor space of 4800 m² (51,700 ft²). The top zone, zone 3, consists of 36 stories with an average floor space of 2000 m² (21,500 ft²). Zones 2 and 3 have sky lobbies at their lowest levels. The sky lobbies are the lobbies for shuttle elevators. They are also designed to meet the requirement for evacuation areas in the event of fire.

The critical design loads for the SSH building were the seismic and wind loads. The response spectra for far-field earthquakes with large magnitudes expected in the Tokyo area appear to have clear peaks around 8 sec. Considering these spectral peaks, a megastructure system with a truss-tube mechanism was employed to keep the SSH building stiff enough to have a fundamental natural period of about 6 sec. This short period helps avoid a lock-in vibration resulting from the vortex shedding in severe winds.

The shore of Tokyo Bay comprises soft soil strata. To overcome the soft soil conditions, special attention has been paid to the foundation system. The proposed foundation system consists of a circular cylindrical wall of a diameter of 162 m (531 ft) with piles and diaphragm walls inside. The thickness of the cylindrical outer wall is 4.0 m (13 ft) in the upper portion. It reaches a depth of 74.5 m (244 ft). This unique foundation system allows this supertall building to be built on such soft soil.

Structural System.   At the site on the shore of the Tokyo Bay area, spectral components of about 8 sec may be pronounced in the response spectra for far-field earthquakes with large magnitudes. Therefore the natural period of 8 sec should be avoided for the SSH building. However, if the fundamental natural period is set to be longer than 8 sec, a lock-in vibration due to strong wind may become a big issue.

Two strategies were established to overcome these problems. The first strategy was to achieve a fundamental natural period of significantly less than 8 sec. The target natural period was set at 6 sec. The second strategy was to select the configuration of the building to minimize the wind loads, especially for the purpose of avoiding a lock-in vibration.

For the first strategy, the structural system selected is a megastructure with a truss tube with steel columns filled with high-strength concrete. This system achieves enough stiffness for the SSH building to have a first natural period of approximately 6 sec.

For the second strategy, the optimum configuration for the SSH building was sought using wind tunnel experiments. Three resolutions were applied to the building. The first resolution was to cut the corners off the building so that the floor plan would become closer to a round shape. The second was to reduce the plan area in the upper zones. The third was to rotate each building zone by 45° with respect to the zone below. This combination effectively broadened the power spectra of wind loads so that lock-in vibration should be unlikely to occur (Fig. 6.11). A perspective of the structural frame is shown in Fig. 6.12.

The soil at the building site is especially soft. To assure enough capacity under this soil condition, a special foundation system has been employed. The unique aspect of the foundation is a continuous circular cylindrical wall system which creates an improved bearing stress distribution and reduces construction cost compared to a conventional system. The continuous outer wall reaches 74.5 m (224 ft) deep. The foundation has a mat slab 5.0 m (16 ft 4 in.) thick between −23.5 and −28.5 m (−77 and −93.5 ft). From the mat slab to the end of the continuous wall, piles and diaphragm walls were used to strengthen the soil contained in the continuous wall. This foundation of a circular cylindrical shape is considered to be rigid enough as a whole.

The thickness of the continuous wall is 4.0 m (13 ft) down to −28.5 m (−93.5 ft). Beyond that depth, the thickness of the wall is kept at 3.2 m (10 ft 6 in.) to the bottom.

Design Criteria.   The design criteria for the SSH building against earthquake and wind loads are as follows:

1. *Under level 1 loads.*   The stresses of the structural main frames should be smaller than the allowable stress. In principle, no uplift is allowed for the foundation.
2. *Under level 2 loads.*   The stresses of the structural main frames should be below the level that can be considered to be elastic as a whole. In addition, no harmful residual deformation due to the foundation movement should be allowed.

Level 1 loads are those that are likely to be experienced by the building during the service period. Level 2 loads are those that can be considered to be the maximum credible loads at the building site.

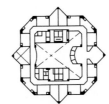

3rd Zone (hotel) typical-floor plan

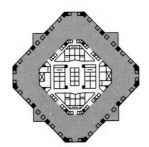

2nd Zone (office) typical-floor plan

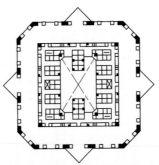

1st Zone (office) typical-floor plan

Fig. 6.11   Typical floor plans; SSH building.

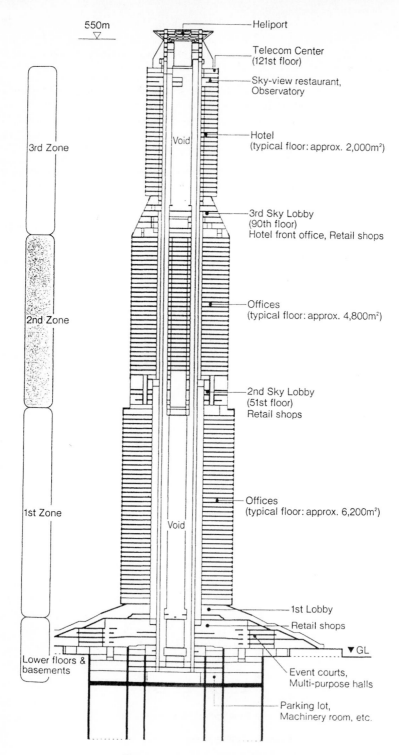

550m ▽

Heliport

Telecom Center
(121st floor)

Sky-view restaurant,
Observatory

Hotel
(typical floor: approx. 2,000m²)

3rd Sky Lobby
(90th floor)
Hotel front office, Retail shops

Offices
(typical floor: approx. 4,800m²)

2nd Sky Lobby
(51st floor)
Retail shops

Offices
(typical floor: approx. 6,200m²)

1st Lobby

Retail shops

▼GL

Event courts,
Multi-purpose halls

Parking lot,
Machinery room, etc.

3rd Zone

2nd Zone

1st Zone

Lower floors &
basements

Void

Void

Fig. 6.12   Elevation; SSH building.

In addition to these design criteria, the discomfort of the building's occupants due to the vibration was assessed for wind and earthquake loads expected to occur once every year. Four hybrid mass dampers (HMDs) will be installed at the top of the SSH building. The weight of each HMD is about 200 tonnes (224 tons). Two HMDs of 100 tonnes (112 tons) have already been installed in a 50-story building in Osaka, Japan.

6.4 CONDENSED REFERENCES/BIBLIOGRAPHY

AISC 1991, *The World's Tallest Building—The Miglin-Beitler Tower*

Architectural Record 1985, *William LeMessurier's Super-Tall Structures*

ASCE 1991, *Building Aims to Be World's Tallest at 1,999 ft*

Engineered Concrete Structures 1990, *The World's Tallest Building—Chicago's Miglin-Beitler Tower*

Mita 1993, *Soil-Structure Interaction Effects on the 121-Story SSH Building*

Watabe 1993a, *Structural Design and Analysis of the 121-Story SSH Building*

Watabe 1993b, *Emerging Needs for Damping Augmenting Systems Applicable to Super Tall Buildings*

Current Questions, Problems, and Research Needs

1. What are the structural systems and building data for other significant high-rise buildings in Europe, South America, and Africa?
2. What is the appropriate way to classify tall building lateral load resisting systems? How are innovative and evolving systems placed within the classification scheme such that cataloging and data collection of structural systems can be continuously updated and of use to the practicing engineer?
3. How are structural schemes tailored to local geographic conditions to produce economical designs?
4. Should there be a professional consensus regarding the acceptability of tall building structures with respect to serviceability issues such as lateral drift, floor vibration, occupant comfort, and floor levelness?
5. What possible structural forms for extra building support such as guyed towers are possible for ultra-tall high-rise buildings? What are the sociological, planning, and maintenance implications for such buildings? What systems are envisioned for the next generation of tall buildings over 600 m (2000 ft) in height?
6. What unique problems are encountered when exposing tall building structural frames on the building perimeter? What are the solutions?
7. What are the structural systems for the future in areas of high seismicity?

Nomenclature

A36. Structural steel with yield strength of 250 MPa (36,000 psi), per ASTM standard.

A572 grade 50. Structural steel with yield strength of 350 MPa (50,000 psi), per ASTM standard.

Acceleration. Rate of change in velocity as a building sways due to wind or earthquake forces.

Allowable stress design or working stress design. Method of proportioning structures such that the computed elastic stress does not exceed a specified limiting stress.

Band beams. Wide, shallow beams used to achieve minimum structural floor depth. A typical size would be 350 mm (13.8 in.) deep by 1500 mm (59 in.) wide.

Basic wind velocity. Wind speed used for design before adjusting for shielding, height, etc. (usually the velocity 10 m (32.8 ft) above ground in smooth, level terrain without significant obstructions).

Bay window. Window projecting from the wall between columns or buttresses.

Beam link. Beam segment between braces, or between a brace and a column.

Bent. Plant framework of beam or truss members that support a floor or roof and the columns that support these members.

Braced frame. Usually a frame which derives its stability primarily from truss action. Most elements have pinned ends and do not develop bending resistance. (These frames usually develop minor bending forces.)

Building standard. Document defining minimum standards for design.

Bundled tube. Structural system in which structural framed tubes are arranged or bundled together so that common walls of contiguous tubes are combined into single walls, thereby forcing compatibility of stresses at the interface of such contiguous tubes. In a bundled tube, individual tube elements may be terminated at any appropriate level.

Castellated beam. Beam fabricated by cutting through the web of the beam with a profile burning machine, separating the two halves, moving one half along the other until the "teeth" of the castellations coincide, and tack welding the two halves together. Deep penetration welding is then used to weld both sides of the web.

Center length. Distance along one member between intersections of centerlines of perpendicular members.

Central business district. Key commercial area inside most modern U.S. cities.

Central-services core. Zone of a high-rise building, often located centrally in plan, where elevators, stairs, toilets, and services shafts are located. Core may be enclosed by concrete walls or steel frames with lightweight cladding.

Chevron.   Inverted V in appearance.

Code.   Building code, a legal document providing design criteria for buildings in a particular jurisdiction.

Coefficient of variation.   Ratio of the standard deviation to the mean of a random variable.

Concentrically braced frame.   Frame in which resistance to lateral load or frame instability is provided by diagonal K or other auxiliary system of bracing.

Core.   Portion of a building that includes elevators, stairs, mechanical shaft, and toilets, often centrally located.

Creep.   Slow time-dependent change in dimensions of concrete under a sustained load, primarily in the direction in which the load acts; a dimensionless quantity having units of strain.

Damping.   Dissipation of energy for dynamic loading.

Dapped girders.   Girders (or beams) having a notch at one or both ends in the underside to accommodate a corbel support within the girder depth or to create additional space for air ducts and the like.

Dead load.   Actual weight of structural elements. (This is a gravity load.)

Differential.   Difference or change between two values.

Doubler.   Plate welded to or parallel to a web or flange to add strength.

Drift.   Lateral displacement due to lateral force.

Ductility.   Ability of a material to absorb energy through deformation without failure.

Eccentrically braced frame.   Frame in which the centerlines of braces are offset from the points of intersection of the centerlines of beams and columns.

Environmental loads.   Loads on a structure due to wind, snow, earthquake, or temperature.

Facade.   Face, especially the principal elevation, of a building.

Factor of safety.   Ratio of the ultimate strength (or yield point) of a material to the working stress assumed in design (stress factor of safety); or ratio of the ultimate load, moment, or shear of a structural member to the working load, moment, or shear, respectively, assumed in design (load factor of safety).

Failure.   Condition where a limit state is reached. This may or may not involve collapse or other catastrophic occurrences.

Fin.   Plate projecting from a member.

Flange moment connection.   Moment connection in which the beam is connected to the flange of the column.

Floor area ratio (FAR).   Specified ratio of permissible floor space to lot area, in which the inducement to reduce lot coverage is an important component. The basic ratio is frequently modified by providing "bonus" or "premium" floor space for such aspects as arcades, setbacks, and plazas. Also called *plot ratio.*

Framed tube.   Perimeter equivalent tube consisting of closely spaced columns and spandrels.

Fundamental period.   Period of the first mode of vibration of a building. (The time taken for the building to sway from its position of maximum deflection on one side of the vertical to its maximum deflection on the other side and back to the first again.)

Hat truss.   Stiff structural trusswork extending from core to perimeter at top of building.

Hybrid building frame.   Frame construction composed of different structural building materials, such as concrete and steel.

Limit states.   Condition in which a structure or a part thereof ceases to fulfill one of its functions or to satisfy the conditions for which it was designed. Limit states can be classified in two categories: (1) *ultimate* limit states, corresponding to the load-carrying capacity of the structure (safety is usually related to these types of limit state), and (2) *serviceability* limit states, related to the criteria governing normal use of the structure.

Limit-state design. Design process that involves identification of all potential modes of failure (limit states) and maintaining an acceptable level of safety against their occurrence. The safety level is usually established on a probabilistic basis.

Load and resistance factor design. Design method in which, at a chosen limit state, load effects and resistances are separately multiplied by factors that account for the inherent uncertainties in the determination of these quantities.

Load combinations. Loads likely to act simultaneously.

Load effects. Moments, shears, and axial forces in a member due to loads or other actions.

Load factors. Factors applied to a load to express probability of not being exceeded; safety factors.

Longitudinal. Direction of the longer plan dimension.

Maximum load (ultimate load). Plastic limit load or stability limit load, as defined; also maximum load-carrying capacity of a structure under test.

Mean recurrence interval (MRI). Average time between occurrences of a random variable that exceed its MRI value. The probability that the MRI value will be exceeded in any occurrence is 1/MRI.

Meandering shear wall. Shear wall following an irregular line in plan. (Not a rectilinear assemblage of walls.)

Medium-rise building. Multistory building neither particularly high nor low; usually in the range of 10 to 20 stories.

Modular tubes. Contiguous framed tubular structural systems which fit together to form a complete bundled tube structure.

Moment resisting frame. Integrated system of structural elements possessing continuity and hence capable of resisting bending forces. (These frames usually develop minor axial forces.)

Mullion. Horizontal or vertical member of a window-wall or curtain-wall system that is normally attached to the floor slab or beams and supports the glass and/or elements of a window wall.

Neoprene. Synthetic rubber having physical properties closely resembling those of natural rubber but not requiring sulfur for vulcanization. It is made by polymerizing chloroprenes, and the latter is produced from acetylene and hydrogen chloride.

Node. Point at which subsidiary parts originate or center.

Nominal load effect. Calculated using a nominal load. The nominal load frequently is defined with reference to a probability level; for example, 50-year mean recurrence interval wind speed used in calculating wind load.

Nominal resistance. Calculated using nominal material and cross-sectional properties and a rationally developed formula based on an analytical and/or experimental model of limit-state behavior.

Outrigger. Stiff structural truss work extending from core to perimeter at any point to distribute column loads between them.

Outstanding. Projecting from main plane.

P-delta effect. Secondary effect of column axial loads and lateral deflection on moments in members.

Probabilistic design. Design method that explicitly utilizes probability theory in the safety checking process.

Probability distribution. Mathematical law that describes the probability that a random variable will assume certain values; either a cumulative distribution function (cdf) or a probability density function is used.

Probability of failure. Probability that the limit state is exceeded or violated.

Probability of survival.   One minus the probability of failure.

Rack.   To deform a rectangle in shear by displacing one side laterally relative to the opposite side.

Resistance.   Maximum load-carrying capacity as defined by a limit state.

Resistance factor.   Partial safety factor to account for the probability of understrength of materials or structural members.

Seismic.   Pertaining to earthquakes.

Shear stud.   Short mild-steel rod with flattened head, welded to a steel member, to transfer shear force between steel and surrounding concrete.

Skewed.   Not parallel or perpendicular.

Slab-type high-rise building.   Building in the shape of a vertical slab standing on the ground on its short dimension.

Spandrel.   Beam spanning between columns on the exterior of a building.

Spandrel beam.   Floor-level beam in the face of a building, usually supporting the edges of the floor slabs.

Staggered truss system.   Structural system for a building with unbraced frames in one direction and frames braced in the other direction by use of story-deep trusses staggered in location at alternate frames on every other floor of the building.

Stocky.   Heavy and thick, composed of elements with low width-to-thickness ratios.

Stressed skin.   Material used for strength and stiffness in its own plane, as in a membrane.

Stub girder.   Vierendeel floor girder comprising the concrete floor as the top chord, a wide-flange beam or column section as the bottom chord, with the chords connected by the floor beams and short lengths of the floor beam (stubs) fixed in line with the bottom chord.

Table forms.   Prefabricated beam and slab formwork complete with vertical props.

Tiebacks.   Mechanical devices for supporting sheeting, consisting of posttensioned rods extending to anchor points in the soil surrounding the excavation or to rock.

Transverse.   Direction of the shorter plan dimension.

Trussed tube system.   Tubular system for tall buildings in which lateral forces are resisted by truss action.

Tube.   Structure with continuous perimeter frame designed to act in a manner similar to that of a hollow cylinder.

Tune.   Adjust carefully.

Unclad.   Not covered by facade.

Vierendeel action.   Using a planar rectangular grid of members working in flexure to act as a truss for longer spans for loads in that plane.

W14.   Nominally 356-mm (14-in.)-deep steel section with wide flange or wide I shape.

Web moment connection.   Moment connection in which beam is connected to web of column.

SYMBOLS

ζ	= critical damping ratio
ρ	= air density
ρ_S	= building density
$\sigma_{\ddot{x}}$	= standard deviation of acceleration in horizontal plane

b = width of building normal to wind direction

C_{FS} = force spectrum coefficient

d = depth of building

E = longitudinal turbulence spectrum; $= 0.47N/(2+N^2)^{5/6}$

g = peak factor; for normally distributed process, $= \sqrt{2 \ln Tn}$

G_{res} = gust factor for resonant component, $= g2(\sigma_v/\overline{V})_h\sqrt{SE/\zeta}$

h = height of building

$H^2(n)$ = mechanical admittance; $= \dfrac{1}{[1 - (n/n_0)^2]^2 + 4\zeta^2(n/n_0)^2}$

L_h = measure of turbulence length scale; $= 1000\,(h/10)^{0.25}$

m = modal mass

$\overline{M}$ = mean base overturning moment; for a square building, can be approximated by $0.6\,(1/2)\rho\overline{V}_h^{\,2}bh^2$

M_I = inertial base bending moment for unit displacement at top of building; for constant density and linear mode shape, $= (1/3)\rho bdh^2(2\pi n_0)^2$

n = frequency of oscillation with an approximately normal distribution

N = reduced frequency; $= nL_h/\overline{V}_h$

n_0 = first-bending-mode natural frequency; can be approximated by $46/h$, where h is height in meters

R = return period, years

S = size factor; $= \dfrac{1}{(1 + 3.5n_0h/\overline{V}_h)(1 + 4n_0b/\overline{V}_h)}$

$S_y(n)$ = spectrum of cross-wind displacement at top of building

$(\sigma_v/\overline{V})_h$ = longitudinal turbulence intensity at height h

T = period under consideration, seconds; usually 600 sec for acceleration criteria

$\overline{V}_h$ = hourly mean wind speed at height h

ABBREVIATIONS

ACI	American Concrete Institute
AISC	American Institute of Steel Construction
ASCE	American Society of Civil Engineers
ASTM	American Society for Testing and Materials
CBF	Concentric braced frame
CCD	Chicago City datum
CTBUH	Council on Tall Buildings and Urban Habitat
EBF	Eccentric braced frame
ECCS	European Convention for Constructional Steelwork

HiRC	High-rise reinforced concrete
HMD	Hybrid mass damper
IISI	International Iron and Steel Institute
IMRF	Intermediate moment-resisting frame
ISO	International Standards Organization
PPR	Partial prestress ratio
SMRF	Special moment resisting frame
TMD	Tuned mass damper
UBC	Uniform Building Code

UNITS

In the table below are given conversion factors for commonly used units. The numerical values have been rounded off to the values shown. The British (Imperial) System of units is the same as the American System except where noted. Le Système International d'Unités (abbreviated "SI") is the name formally given in 1960 to the system of units partly derived from, and replacing, the old metric system.

SI	American	Old metric
	Length	
1 mm	0.03937 in.	1 mm
1 m	3.28083 ft	1 m
	1.093613 yd	
1 km	0.62137 mi	1 km
	Area	
1 mm^2	0.00155 in.^{2}	1 mm^2
1 m^2	10.76392 ft^2	1 m^2
	1.19599 yd^2	
1 km^2	247.1043 acres	1 km^2
1 hectare	2.471 acres[1]	1 hectare
	Volume	
1 cm^3	0.061023 in.^{3}	1 cc
		1 ml
1 m^3	35.3147 ft^3	1 m^3
	1.30795 yd^3	
	264.172 gal[2] liquid	
	Velocity	
1 m/sec	3.28084 ft/sec	1 m/sec
1 km/hr	0.62137 mi/hr	1 km/hr

SI	American	Old metric
	Acceleration	
1 m/sec^2	3.28084 ft/sec^2	1 m/sec^2
	Mass	
1 g	0.035274 oz	1 g
1 kg	2.2046216 lb[3]	1 kg
	Density	
1 kg/m^3	0.062428 lb/ft^3	1 kg/m^3
	Force, weight	
1 N	0.224809 lbf	0.101972 kgf
1 kN	0.1124045 ton[4]	
1 MN	224.809 kips	
1 kN/m	0.06853 kips/ft	
1 kN/m^2	20.9 lbf/ft^2	
	Torque, bending moment	
1 N-m	0.73756 lbf-ft	0.101972 kgf-m
1 kN-m	0.73756 kip-ft	101.972 kgf-m
	Pressure, stress	
1 N/m^2 = 1 Pa	0.000145038 psi	0.101972 kgf/m^2
1 kN/m^2 = 1 kPa	20.8855 psf	
1 MN/m^2 = 1 MPa	0.145038 ksi	
	Viscosity (dynamic)	
1 N-sec/m^2	0.0208854 lbf-sec/ft^2	0.101972 kgf-sec/m^2
	Viscosity (kinematic)	
1 m^2/sec	10.7639 ft^2/sec	1 m^2/sec
	Energy, work	
1 J = 1 N-m	0.737562 lbf-ft	0.00027778 W-hr
1 MJ	0.37251 hp-hr	0.27778 kW-hr
	Power	
1 W = 1 J/sec	0.737562 lbf ft/sec	1 W
1 kW	1.34102 hp	1 kW
	Temperature	
K = 273.15 + °C	°F = (°C × 1.8) + 32	°C = (°F − 32)/1.8
K = 273.15 + 5/9(°F − 32)		
K = 273.15 + 5/9(°R − 491.69)		

(1) Hectare as an alternative for km^2 is restricted to land and water areas.
(2) 1 m^3 = 219.9693 Imperial gallons.
(3) 1 kg = 0.068522 slugs.
(4) 1 American ton = 2000 lb.   1 kN = 0.1003612 Imperial ton.   1 Imperial ton = 2240 lb.

Abbreviations for Units

Btu	British thermal unit	kW	kilowatt
°C	degree Celsius (centigrade)	lb	pound
cm^3	cubic centimeters	lbf	pound force
cm	centimeter	lb_m	pound mass
°F	degree Fahrenheit	MJ	megajoule
ft	foot	MPa	megapascal
g	gram	m	meter
gal	gallon	mi	mile
hp	horsepower	ml	milliliter
hr	hour	mm	millimeter
Imp	British Imperial	MN	meganewton
in.	inch	N	newton
J	joule	oz	ounce
K	kelvin	Pa	pascal
kg	kilogram	psf	pounds per square foot
kgf	kilogram-force	psi	pounds per square inch
kip	1000 pound force	°R	degree Rankine
km	kilometer	sec	second
kN	kilonewton	slug	14.594 kg
kPa	kilopascal	W	watt
ksi	kips per square inch	yd	yard

References/Bibliography

AISC, 1983
  MODERN STEEL CONSTRUCTION, American Institute of Steel Construction, Chicago, Ill., 2d Quarter.

AISC, 1987
  ONE LIBERTY PLACE—EFFICIENCY AND ELEGANCE IN THE CRADLE OF HISTORY, *Modern Steel Construction,* no. 2, pp. 9–14.

AISC, 1991
  THE WORLD'S TALLEST BUILDING—THE MIGLIN-BEITLER TOWER, *Modern Steel Construction,* August.

Architectural Record, 1985
  WILLIAM LEMESSURIER'S SUPER-TALL STRUCTURES, *Architectural Record,* January/February.

Architecture, 1988
  EXPLORING COMPOSITE STRUCTURES, *Architecture,* March.

Architecture, 1988
  TWO UNION SQUARE, *Architecture,* March.

Architecture, 1990
  HIGH STRENGTH, *Architecture,* October.

Architecture and Urbanism, 1991
  TWO UNION SQUARE, *Architecture and Urbanism,* February.

ASCE, 1986
  COMPUTER CUTS TOWER STEEL, *Civil Engineering,* March.

ASCE, 1990
  AUSSIE STEEL, *Civil Engineering,* December.

ASCE, 1991
  BUILDING AIMS TO BE WORLD'S TALLEST AT 1,999 FT., *Civil Engineering,* March.

Assefpour-Dezfuly, M., Hugaas, B. A., and Browrigg, A., 1990
  FIRE RESISTANT HIGH STRENGTH LOW ALLOY STEELS, *Materials Science and Technology,* vol. 6, December.

Australia Post Publ., 1988
  CHIFLEY SQUARE ON THE MOVE STRUCTURES, no. VBP 8810.

Beck, V., 1991
  FIRE SAFETY SYSTEMS DESIGN USING RISK ASSESSMENT MODELS—DEVELOPMENTS IN AUSTRALIA, *Fire Safety Science,* Proceedings of the 3d International Symposium, Elsevier.

Bennetts, I. D., Almand, K. H., Thomas, I. R., Proe, D. J., and Lewins, R. R., 1989
  FIRE IN CARPARKS, BHP Melbourne Research Laboratories, Australia, Report MRL/PS69/85/005, August.

Bennetts, I. D., Proe, D. J., Lewins, R. R., and Thomas, I. R., 1985
  OPEN-DECK CARPARK FIRE TESTS, BHP Melbourne Research Laboratories, Australia, Report MRL/PS69/85/001.

Bond, G. V. L., 1975
  FIRE AND STEEL CONSTRUCTION: WATER COOLED HOLLOW COLUMNS, *Constrado.*

British Steel, 1992
DESIGN MANUAL FOR CONCRETE FILLED COLUMNS—PART 1: STRUCTURAL DE-SIGN/PART 2: FIRE-RESISTANT DESIGN.

Brozetti, J., Law, M., Pettersson, O., and Witteveen, J., 1983
FIRE PROTECTION OF STEEL STRUCTURES—EXAMPLES OF APPLICATIONS, IABSE Periodical 2, May.

Building, 1990
DOUBLE STRENGTH, *Building,* July.

Building Design and Construction, 1984
BUILDING DESIGN AND CONSTRUCTION, Cahners Publishing, June.

Chen, P. W., and Robertson, K. E., 1973
HUMAN PERCEPTION THRESHOLDS TO HORIZONTAL MOTION, *Journal of the Structural Division, ASCE,* vol. 98, pp. 1681–1695.

Civil Engineer, 1987
CONCRETE STRENGTH RECORD JUMPS 36%, *Civil Engineer,* October.

Concrete Today, 1989
ALWAYS SOMETHING NEW IN CONCRETE, *Concrete Today,* Spring.

Construction Specifier, 1988
INNOVATIVE COMPOSITE CONSTRUCTION, *Construction Specifier,* April.

Construction Steel, 1990
THE MANY FACES OF THE BOND BUILDING, *Construction Steel,* February.

CTBUH, Group CL, 1980
FIRE, chapter CL-4, *Tall Building Criteria and Loading,* vol. CL of *Monograph on Planning and Design of Tall Buildings,* ASCE, New York.

CTBUH, Group SC, 1980
TALL BUILDING SYSTEMS AND CONCEPTS, vol. SC of *Monograph on Planning and Design of Tall Buildings,* ASCE, New York.

CTBUH, Committee 8A, 1992
FIRE SAFETY IN TALL BUILDINGS, McGraw-Hill, New York.

Davenport, A. G., 1967
GUST LOADING FACTORS, *Journal of the Structural Division, ASCE,* vol. 93, no. ST3.

Drew, R. J., and St. Claire Johnson, C., 1990
RIALTO TOWERS PROJECT SEISMIC RESPONSE ANALYSIS AND EVALUATION, vols. 1, 2, 3, and 4, June.

ECCS, 1988
CALCULATION OF THE FIRE RESISTANCE OF CENTRALLY LOADED COMPOSITE STEEL-CONCRETE COLUMNS TO THE STANDARD FIRE, Technical Note no. 55, European Convention for Constructional Steelwork, Brussels, Belgium.

Engineered Concrete Structures, 1990
THE WORLD'S TALLEST BUILDING—CHICAGO'S MIGLIN-BEITLER TOWER, vol. 3, no. 3, December.

Engineering News Record, 1988
SYDNEY SKYSCRAPER SETS SAIL, *Engineering News Record,* August 11.

Engineering News Record, 1989
19,000 PSI, *Engineering News Record,* February.

Engineering News Record, 1990
INNOVATIVE TECHNIQUES, *Engineering News Record,* April.

Engineering News Record, 1991
SYDNEY TOWER TESTS AUSTRALIANS, *Engineering News Record,* June 17.

Falconer, D., and Beedle, L. S., 1984
CLASSIFICATION OF TALL BUILDING SYSTEMS, Council Report no. 442.3, Council on Tall Buildings and Urban Habitat, Bethlehem, Pa.

George, S. F., 1990
WELLINGTON'S WINDS SHAPED THE CAPITAL'S TALLEST BUILDING, *New Zealand Engineering,* September.

Gillespie, B. J., Nasim, S., and St. Claire Johnson, C., 1990
DESIGN AND CONSTRUCTION OF STEEL FRAMED HIGHRISE BUILDINGS, Proceedings of Seminar on Steel Structures, Singapore.

Grossman, J. S., 1985
780 THIRD AVENUE, THE FIRST HIGH-RISE DIAGONALLY BRACED CONCRETE STRUCTURE, *Concrete International, Design and Construction,* vol. 7, no. 2, February, pp. 53–56.

Grossman, J. S., 1989
SLENDER STRUCTURES—THE NEW EDGE (II), Proceedings of the International Conference on Tall Buildings and City Development, Brisbane, Australia, October, pp. 93–99.

Grossman, J. S., 1990
SLENDER CONCRETE STRUCTURES—THE NEW EDGE, *ACI Structural Journal,* vol. 87, no. 1, January-February, pp. 39–52.

Grossman, J. S., Cruvellier, M. R., and Stafford-Smith, B., 1986
BEHAVIOR, ANALYSIS AND CONSTRUCTION OF A BRACED-TUBE CONCRETE STRUCTURE, *Concrete International, Design and Construction,* vol. 8, no. 9, September, pp. 32–42.

Holmes, J. D., 1987
MODE SHAPE CORRECTIONS FOR DYNAMIC RESPONSE TO WIND, *Engineering Structures,* vol. 9, pp. 210–212.

Horvilleur, J. F., 1992
DESIGN OF THE NATIONS BANK CORPORATE CENTER, *The Structural Design of Tall Buildings,* Vol. 1, pp. 75–119.

Hose, R. M., 1990
STRUCTURAL DESIGN FOR THE RIALTO TOWERS, Melbourne.

IISI, 1993
FIRE ENGINEERING DESIGN FOR STEEL STRUCTURES: STATE OF THE ART, International Iron and Steel Institute, Brussels, Belgium, 1131.

Irwin, A., 1988
MOTION IN TALL BUILDINGS, Proceedings of the 3d International Conference on Tall Buildings, Second Century of the Skyscraper, Chicago, Ill., Council on Tall Buildings and Urban Habitat, Bethlehem, Pa.

ISO, 1985
FIRE-RESISTANCE TESTS—ELEMENTS OF BUILDING CONSTRUCTION, AS 1530, Part 4, International Standards Organization, Geneva.

Itoh, M., 1991
WIND RESISTANT DESIGN OF A TALL BUILDING WITH AN ELLIPSOIDAL CROSS SECTION, Proceedings of the 2d Conference on Tall Buildings in Seismic Regions, Los Angeles Tall Buildings Structural Design Council and Council on Tall Buildings and Urban Habitat, May.

Iyengar, H., 1992
HOTEL DE LAS ARTES TOWER, BARCELONA, SPAIN, *Structural Engineering International,* vol. 2, no. 3, August.

Journal of Wind Engineering and Industrial Aerodynamics, 1990
OPTIMIZATION OF TALL BUILDINGS FOR WIND LOADING, Elsevier.

Khan, F. R., 1966
OPTIMIZATION OF BUILDING STRUCTURES, Proceedings of Structural Engineering Conference held at University of Illinois, May.

Kilmister, M. B., 1983
DESIGN AND CONSTRUCTION OF THE LUTH HEADQUARTERS BUILDING, KUALA
LUMPUR, Proceedings of the Biennial Conference of the Concrete Institute of Australia,
June.

Kruppa, J., 1981
FIRE-RESISTANCE OF EXTERNAL STEEL COLUMNS, Final Report, Technical Steel Research, Commission of the European Communities.

Kruppa, J., Schaumann, P., Schleich, J. B., and Twilt, L., 1990
STRUCTURAL FIRE DESIGN, Part 10, draft Eurocode 4, Commission of the European Communities, April.

Kurzeme, M., Hose, R. M., and Nasim, S., 1990
THE OUB CENTRE TOWER FOUNDATIONS, SINGAPORE, Proceedings of the Conference
on Deep Foundation Practice, Singapore.

Kurzeme, M., and Rush, M. C., 1985
DEEP CAISSON FOUNDATIONS FOR OUB CENTRE, SINGAPORE, Proceedings of the 8th
South East Asian Geotechnical Conference, Kuala Lumpur.

Law, M., and O'Brien, T., 1981
FIRE SAFETY OF EXTERNAL STEELWORK, *Constrado.*

L'Indústria Italiàna del Cemènto, 1987
THE LUTH BUILDING IN KUALA LUMPUR (MALAYSIA), *L'Indústria Italiàna del
Cemènto,* no. 613, July-August, pp. 472–485.

Martin, O., and Peyton, J., 1989
WIND DESIGN OF FOUR BUILDINGS UP TO 306 m TALL, *Reinforced Concrete Digest,*
March.

Maruoka, Y., Tsubaki, H., and Hisatoku, T., 1992
DEVELOPMENT AND TEST RESULTS OF SM520B-NFR FIRE RESISTANT STEEL FOR
PROCTER & GAMBLE FAR EAST, INC. JAPAN HEADQUARTERS BUILDING, Pro-
ceedings of the Pacific Structural Steel Conference, Tokyo.

McBean, P. C., 1990
THE MYER CENTRE, ADELAIDE—A CASE STUDY, Proceedings of the Structural Engineering Conference, Adelaide, The Institution of Engineers, Australia.

Meinhardt, W. L., and Nasim, S., 1990
THE OUB CENTRE—QUALITY DELIVERY, Proceedings of Zero Defects Construction
[ap]90 Quality Delivery, Singapore.

Meinhardt, W. L., and Nisbet, R. D., 1984
SUPERSTRUCTURE DESIGN FOR THE OVERSEAS UNION BANK BUILDING, SINGA-
PORE, Proceedings of the International Conference on Tall Buildings, Singapore.

Melbourne, W. H., 1977
PROBABILITY DISTRIBUTIONS ASSOCIATED WITH THE WIND LOADING OF
STRUCTURES, *Civil Engineering Transactions,* The Institution of Engineers, Australia,
vol. CE19, no. 1, pp. 58–67.

Melbourne, W. H., 1980
NOTES AND RECOMMENDATIONS ON ACCELERATION CRITERIA FOR OCCU-
PANCY COMFORT IN TALL STRUCTURES, Private Communications in Commercial
Wind Tunnel Investigation Reports.

Melbourne, W. H., and Cheung, J. C. K., 1988
DESIGNING FOR SERVICEABLE ACCELERATIONS IN TALL BUILDINGS, Proceedings
of the 4th International Conference on Tall Buildings, Hong Kong and Shanghai, pp.
148–155.

Melbourne, W. H., and Nisbet, R. D., 1985
AEROPLASTIC MODEL TESTS AND THEIR APPLICATION FOR THE OUB CENTRE,
SINGAPORE, Proceedings of the International Conference on Testing and Instrumentation
in Building and Construction, Singapore.

Melbourne, W. H., and Palmer, T. R., 1992
ACCELERATIONS AND COMFORT CRITERIA FOR BUILDINGS UNDERGOING COMPLEX MOTIONS, *Journal of Wind Engineering and Industrial Aerodynamics,* vol. 41, pp. 105–116.

Mita, A., and Fuchimoto, M., 1993
SOIL-STRUCTURE INTERACTION EFFECTS ON THE 121-STORY SSH BUILDING, Proceedings of the International Conference on Tall Buildings, Rio de Janeiro, Brazil, May 17–19.

O'Meagher, A. J., Bennetts, I. D., Stevens, L. K., and Hutchinson, G. L., 1993
BEHAVIOR OF COMPOSITE COLUMNS IN FIRE, BHP Melbourne Research Laboratories, Australia, Report BHPR/PPA/R/93/001/SG3C.

Platten, D. A., 1986
POSTMODERN ENGINEERING, *Civil Engineering,* June

Platten, D. A., 1988
MOMENTUM PLACE: STEEL SOLVES COMPLEX GEOMETRIES, *Modern Steel Construction,* Feb., No. 2.

Pettersson, O., Magnusson, S. E., and Thor, J., 1976
FIRE ENGINEERING DESIGN OF STEEL STRUCTURES, Publication no. 50, Swedish Institute of Steel Construction.

Reed, J. W., 1971
WIND INDUCED MOTION AND HUMAN COMFORT, Research Report 71-42, Massachusetts Institute of Technology, Cambridge, Mass.

Sakumoto, Y., Keira, K., Takagi, M., Kaminaga, K., and Gokan, S., 1992
APPLICATION OF FIRE-RESISTANT STEEL TO A HIGH-RISE BUILDING, Proceedings of the Pacific Structural Steel Conference, Tokyo.

Saunders, J. W., and Melbourne, W. H., 1975
TALL RECTANGULAR BUILDING RESPONSE TO CROSS-WIND EXCITATION, Proceedings of the 4th International Conference on Wind Effects on Buildings and Structures, London, Cambridge University Press, pp. 369–380.

Taranath, B. S., 1988
STRUCTURAL ANALYSIS AND DESIGN OF TALL BUILDINGS, McGraw-Hill, New York.

Thomas, I. R., Almand, K. H., Bennetts, I. D., Proe, D. J., and Lewins, R. R., 1989
FIRE IN MIXED OCCUPANCY BUILDINGS, BHP Melbourne Research Laboratories, Australia, Report MRL/PS69/89/004, August.

Thomas, I. R., Bennetts, I. D., Proe, D. J., and Lewins, R. R., 1992a
FIRE TESTS OF THE 140 WILLIAM STREET OFFICE BUILDINGS, BHP Melbourne Research Laboratories, Australia, Report BHPR/ENG/R/92/043/SG2C, January.

Thomas, I. R., Bennetts, I. D., Proe, D. J., and Lewins, R. R., 1992b
THE EFFECT OF FIRE ON 140 WILLIAM STREET—A RISK ASSESSMENT, BHP Melbourne Research Laboratories, Australia, Report BHPR/ENG/R/92/044/SG2C, January.

Vickery, B. J., 1966
ON THE ASSESSMENT OF WIND EFFECTS ON ELASTIC STRUCTURES, *Civil Engineering Transactions,* The Institution of Engineers, Australia, pp. 183–192.

Vickery, B. J., 1969
ON THE RELIABILITY OF GUST LOADING FACTORS, U.S. Department of Commerce, National Bureau of Standards Building Science ser. 30.

Watabe, M., and Mita, A., 1993a
STRUCTURAL DESIGN AND ANALYSIS OF THE 121-STORY SSH BUILDING, Proceedings of the International Conference on Tall Buildings, Rio de Janeiro, Brazil, May 17–19.

Watabe, M., and Mita, A., 1993b
EMERGING NEEDS FOR DAMPING AUGMENTING SYSTEMS APPLICABLE TO SUPER TALL BUILDINGS, Proceedings of the International Workshop on Structural Control, Honolulu, Hawaii, August 5–7.

Watson, K. B., and O'Brien, L. J., 1990
    TUBULAR COMPOSITE COLUMNS AND THEIR DEVELOPMENT IN AUSTRALIA, Pro-
    ceedings of the Structural Engineering Conference, Adelaide, The Institution of Engineers,
    Australia.
Wyett, G. W., and Bennetts, I. D., 1987
    STRUCTURAL FIRE ENGINEERING IN BUILDING DESIGN—A CASE STUDY, Proceed-
    ings of the First National Structural Engineering Conference, The Institution of Engineers,
    Australia.

Contributors

The following is a list of those who have contributed their time and effort to make this volume possible. The names, affiliations, cities, and countries of each contributor are given.

Ian D. Bennetts, BHP Melbourne Laboratories, Melbourne, Australia

Joseph Burns, LeMessurier Consultants, Inc., Chicago, Illinois, USA

Brian Cavill, VSL Prestressing (Aust) Pty. Ltd., Sydney, Australia

Joseph P. Colaco, CBM Engineers, Houston, Texas, USA

Henry J. Cowan, University of Sydney, Sydney, Australia

P. H. Dayawansa BHP Melbourne Laboratories, Melbourne, Australia.

James G. Forbes, Irwin Johnston and Partners, Sydney, Australia

Eiji Fukuzawa, Kajima Design, Tokyo, Japan

Max B. Kilmister, Connell Wagner Consulting Engineers, Brisbane, Australia

Ryszard M. Kowalczyk, Department of Civil Engineering, University of Beira Interior, Covilha, Portugal (former: Bialystok University of Technology, Bialystok, Poland)

Owen Martin, Connell Wagner Rankine and Hill, Sydney, Australia

William Melbourne, Department of Mechanical Engineering, Monash University, Melbourne, Australia

Seiichi Muramatsu, Kajima Design, Tokyo, Japan

T. Okoshi, Nihon Sekkei, Tokyo, Japan

Ahmad Rahimian, The Office of Irwin G. Cantor, New York, New York, USA

Thomas Scarangello, Thornton-Tomasetti Engineers, New York, New York, USA

Robert Sinn, Skidmore Owings and Merrill, Chicago, Illinois, USA

Richard Tomasetti, Thornton-Tomasetti Engineers, New York, New York, USA

A. Yamaki, Nihon Sekkei, Tokyo, Japan

Building Index

Pictures of the buildings appear on the italicized pages.

Name Index

Subject Index